Holt Mathematics

Chapter 6 Resource Book

HOLT, RINEHART AND WINSTON
A Harcourt Education Company
Orlando • Austin • New York • San Diego • London

Printed in the United States of America

ISBN 0-03-078222-8

5 6 170 09 08

CONTENTS

Blackline Masters

Date ________

Dear Family,

In this chapter, your child will learn how to collect, analyze, present and interpret data in order to solve real world problems. Advertisers present information to children of all ages. The ability to evaluate this information in order to make informed decisions is a skill your child will use throughout his or her life.

Data can be organized in various ways so that patterns and relationships are clearly visible. In one example, your child will be asked to use temperature data to make a table. The table can then be used to find a pattern in the data and draw conclusions.

At 10:00 A.M., the temperature was 62°F. At noon, it was 65°F. At 2:00 P.M., it was 68°F. At 4:00 P.M., it was 70°F. At 6:00 P.M., it was 66°F.

Time	**Temperature** (°F)
10:00 A.M.	62
12:00 noon	65
2:00 P.M.	68
4:00 P.M.	70
6:00 P.M.	66

Your child will be asked to find the range, mean, median, and mode of a set of data. These are important for analyzing the values at the center of the data.

Heights of Vertical Jumps (in.)						
13	23	21	20	21	24	18

First, the data is written in numerical order.

13, 18, 20, 21, 21, 23, 24

Range: $24 - 13 = 11$ *Subtract the least value from the greatest value.*

Mean: $13 + 18 + 20 + 21 + 21 + 23 + 24 = 140$ *Add all values.*

$140 \div 7 = 20$ *Divide the sum by the number of items.*

Median: 21 *There is an odd number of items, so find the middle value.*

Mode: 21 *21 occurs most often.*

The range is 11 in.; the mean is 20 in.; the median is 21 in.; and the mode is 21 in.

Your child will also learn to display and analyze data using graphs. One type of graph introduced is the bar graph.

The bar graph can be used to answer questions such as:

Which has the most rainfall?

Find the highest bar.

The rain forest has the most rainfall.

Which have an average yearly rainfall greater than 80 inches?

Find the bars whose heights measure greater than 80.

The savanna, the rain forest, and the deciduous forest have average yearly rainfalls greater than 80 inches.

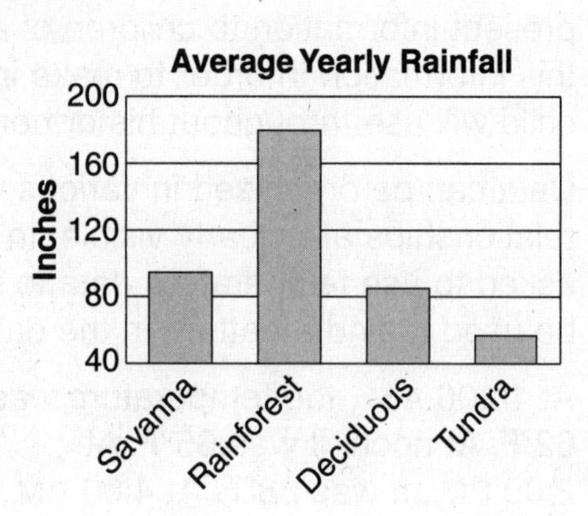

Cities, towns, and neighborhoods are often laid out on a grid to make it easier to map and find locations. Your child will learn to use an **ordered pair** of numbers to locate specific points on a **coordinate plane,** as in this example:

You can use the coordinate plane to name the ordered pair for each location.

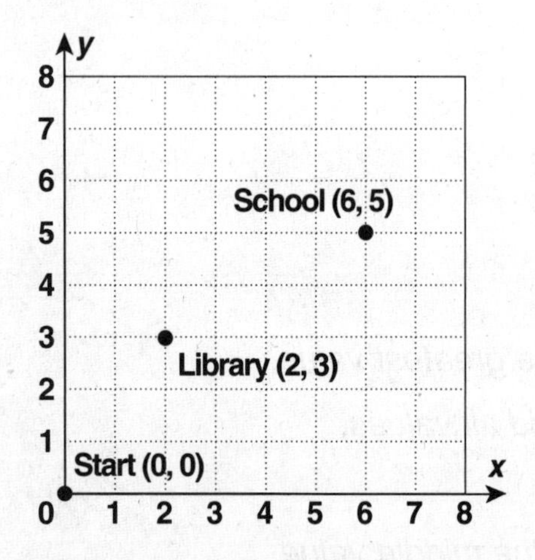

To find the library:
Start at (0, 0). Move right 2 units and then up 3 units.
The library is located at (2, 3).

To find the school:
Start at (0, 0) Move right 6 units and then up 5 units.
The school is located at (6, 5).

For additional resources, visit go.hrw.com and enter the keyword MR7 Parent.

Name ______________________ Date __________ Class __________

LESSON 6-1

Practice A

Problem Solving Skill: Make a Table

Complete each activity and answer the questions.

1. There are five children in the Cooper family. July is 8 years old, Brent is 16 years old, Michael is 10 years old, Andrea is 14 years old, and Matthew is 12 years old. Use this data to complete the table at right showing the children's ages in order from youngest to oldest.

Cooper Children Ages

Child	Age

2. What pattern do you see in the table's data?

____

____

3. On Monday, the low temperature was 41°F. On Tuesday, the low temperature was 38°F. On Wednesday, the low temperature was 35°F. On Thursday, the low temperature was 32°F and on Friday the low temperature was 29°F. Use this data to complete the table at right.

Daily Low Temperatures

Day	Temperature (°F)

4. What pattern do you see in the table's data?

____

____

5. Look at the table you completed for Exercise 1. Name a different way that you could organize the same data in the table.

____

____

6. Look at the table you completed for Exercise 3. Name a different way that you could organize the same data in the table.

____

____

Name ______________________ Date __________ Class __________

LESSON 6-1

Practice B

Problem Solving Skill: Make a Table

Complete each activity and answer each question.

1. Pizza Express sells different-sized pizzas. The jumbo pizza has 20 slices. The extra large has 16 slices. The large has 12 slices. There are 8 slices in a medium, and 6 slices in a small. A personal-sized pizza has 4 slices. Use this data to complete the table at right, from largest to smallest pizza.

2. What pattern do you see in the table's data?

____

____

____

3. A plain large pizza at Pizza Express costs $13.75. A large pizza with one topping costs $14.20. A 2-topping large pizza costs $14.65, and a 3-topping large pizza costs $15.10. If you want 4 toppings on your large pizza, it will cost you $15.55. Use this data to complete the table at right.

4. What pattern do you see in the table's data?

____

____

5. How much does each slice of a 1-topping large pizza from Pizza Express cost? Round your answer to the nearest hundredth of a dollar.

____

6. You and three friends buy two large pizzas from Pizza Express. One pizza has pepperoni and onions, and one pizza is plain. If you equally share the total price, how much will you each pay? How many slices will you each get?

____

Name ______________________ Date ___________ Class ___________

LESSON 6-1

Practice C

Problem Solving Skill: Make a Table

Complete each activity and answer each question.

1. At 9:00 A.M., the temperature in Yuma, Arizona, was 79°F. At noon, the temperature was 83°F. At 2:00 P.M., the temperature was 87°F. At 4:00 P.M., the temperature was 84°F. At 6:00 P.M., the temperature was 81°F. By 8:00 P.M., the temperature in Yuma was 78°F. Use this data to complete the two tables below. Each table's data should be organized differently.

Table A

Table B

2. Use Table A to find a pattern in the data and draw a conclusion.

____

____

____

3. Use Table B to find a pattern in the data and draw a conclusion.

____

____

____

4. When would each of these tables be most useful for finding and comparing data?

____

____

____

Name ______________________ Date __________ Class __________

LESSON 6-1

Reteach

Problem Solving Skill: Make a Table

You can make a table to organize data. Then you can use the table to see patterns and draw conclusions.

During the week-long book fair, 324 books were sold. On Monday, 45 books were sold. On Tuesday, students bought 58 books. On Wednesday, 79 books were sold. Sixty-two books were sold on Thursday, and students bought 51 books on Friday.

Day	Books Sold
Monday	45
Tuesday	58
Wednesday	79
Thursday	62
Friday	51

To make a table, arrange the information in order by days so you can see patterns over time. Remember to make headings for each column of the table.

From the table, you can see that the number of books sold increased from Monday to Wednesday, and decreased from Wednesday to Friday.

Use the data to make a table. Then use the table to find a pattern in the data and draw a conclusion.

1. During the championship series, the school basketball team earned 24 points in the first game, 28 points in the second game, 33 points in the third game, 42 points in the fourth game, and 49 points in the last game.

____

____

____

____

2. In the sixth grade, 18 students study Spanish, 35 students study French, 11 students study Latin, and 5 students study no foreign language at all.

____

____

Name ______________________ Date __________ Class __________

LESSON 6-1

Challenge

Liberty Logic

You can use tables and logic to organize information and solve problems. For example, you have some measurements for different parts of the Statue of Liberty, but you do not know which measurement goes with which part. To solve the problem, first organize all the possibilities in a logic table. Then use the clues to fill out the table.

Because each part has only one measurement, there can be only one **YES** in each row and column of your logic table.

Three measurements for parts of the statue's face are 30 inches, 36 inches, and 54 inches. Which of those measurements are for the width of her mouth, the length of her nose, and the width of each of her eyes?

Clue 1: Her mouth is wider than each of her eyes.
Clue 2: The length of her nose is greater than the width of her mouth.

	30 inches	36 inches	54 inches
Width of Mouth			
Length of Nose			
Width of each Eye			

Three measurements for the tablet she holds are 24 inches, 163 inches, and 283 inches. Use the clues and logic table below to find the length, width, and thickness of the Statue of Liberty tablet.

Clue 1: The tablet is longer than it is wide.
Clue 2: The tablet is less than 100 inches thick.

Name ______________________ Date ____________ Class ____________

LESSON 6-1

Problem Solving

Problem Solving Skill: Make a Table

Complete each activity and answer each question.

1. In January, the normal temperature in Atlanta, Georgia, is 41°F. In February, the normal temperature in Atlanta is 45°F. In March, the normal temperature in Atlanta is 54°F, and in April, it is 62°F. Atlanta's normal temperature in May is 69°F. Use this data to complete the table at right.

2. Use your table from Exercise 1 to find a pattern in the data and draw a conclusion about the temperature in June.

____

____

____

3. In what other ways could you organize the data in a table?

____

____

Circle the letter of the correct answer.

4. In which month given does Atlanta have the highest temperature?
   A February
   B March
   C April
   D May

5. In which month given does Atlanta have the lowest temperature?
   F January
   G February
   H March
   J April

6. Which of these statements about Atlanta's temperature data from January to May is true?
   A It is always higher than 40°F.
   B It is always lower than 60°F.
   C It is hotter in March than in April.
   D It is cooler in February than in January.

7. Between which two months in Atlanta does the normal temperature change the most?
   F January and February
   G February and March
   H March and April
   J April and May

Name ______________________ Date __________ Class __________

LESSON 6-1

Reading Strategies

Reading a Table

Jill jumps rope each day as part of her fitness program. She made a **table** to keep track of how much time she spent jumping rope each day for one week.

Day	Time
Monday	15 minutes
Tuesday	18 minutes
Wednesday	18 minutes
Thursday	21 minutes
Friday	21 minutes
Saturday	24 minutes
Sunday	24 minutes

1. What headings are shown in the table?

____

2. What does "time" stand for in the table?

____

3. How long did Jill jump rope on Tuesday?

____

4. What other day of the week did Jill jump rope for the same length of time? _______________________

5. On which day did Jill jump rope for 15 minutes?

____

6. List in order from least to greatest the different times Jill spent jumping rope.

____

7. What pattern do you notice in this list of times?

____

8. If the pattern continues, how much time would you expect Jill to jump rope on the following Monday? _______________________

9. How did the table help you understand Jill's fitness program?

____

____

Name ______________________ Date ____________ Class ____________

LESSON 6-1

Puzzles, Twisters & Teasers

Making the Grade

Charlie, Pat, Oliver, Eduardo, and Hannah each received different grades from B all the way up to an A+ on the last test. Given the following clues, use the yes/no table to determine the grade each of the students received.

1. Oliver received some sort of A.
2. Hannah and Pat each got a higher grade than Eduardo.
3. Charlie got the A−.
4. Pat's grade was higher than Oliver's was.

	A+	A	A−	B+	B
Hannah					
Eduardo					
Oliver			NO	NO	NO
Pat					
Charlie					

To solve the riddle, fill the spaces above each of the grades with the first letter of the name of the person who received that grade.

What do you get when you cross a dog and a hen?

___ ___ ___ ___ ___ ___ D ___ G G S

A+ A A A− B+ B B

Name ______________________ Date __________ Class __________

LESSON 6-2

Practice A

Mean, Median, Mode, and Range

Find the mean of each data set.

1.

Length of Worms (in.)	3	5	4	2	6

____

2.

Ages of Brothers (yr)	12	16	15	14	8

____

Find the mean, median, mode, and range of each data set.

3.

Heights of Trees (m)	7	11	9	7	6

____

4.

Sizes of Bottled Juice (L)	6	12	12	16	24

____

5.

Football Team Wins (games per season)	10	8	10	8	14

____

____

6. Tammy is 14 years old. She has a younger sister and an older brother. Her sister is 12 years old. The mean of all their ages is 14. How old is Tammy's brother?

____

7. The mode of Nevin's four math quiz scores last month is 85 points. On three of the quizzes, he earned the following scores: 90, 86, and 85. What was the score of Nevin's other quiz?

____

Name ______________________ Date ____________ Class ____________

LESSON 6-2

Practice B

Mean, Median, Mode, and Range

Find the mean of each data set.

1.

Brian's Math Test Scores	86	90	93	85	79	92

____

2.

Heights of Basketball Players (in.)	72	75	78	72	73

____

Find the mean, median, mode, and range of each data set.

3.

School Sit-Up Records (sit-ups per minute)	31	28	30	31	30

____

____

4.

Team Heart Rates (beats per min)	70	68	70	72	68	66

____

5.

Daily Winter Temperatures (°F)	45	50	47	52	53	45	51

____

6. Anita has two sisters and three brothers. The mean of all their ages is 6 years. What will their mean age be 10 years from now? Twenty years from now?

____

7. In a class of 28 sixth graders, all but one of the students are 12 years old. Which two data measurements are the same for the student's ages? What are those measurements?

____

Name ________________________________ Date ____________ Class ____________

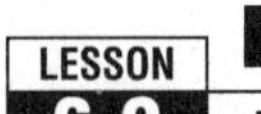

Practice C

Mean, Median, Mode, and Range

Find the mean of the data set.

1.

Monthly Girl Scout Cookie Sales (boxes per person)									
22	13	47	11	8	16	15	14	13	17

____

Find the mean, median, mode, and range of each data set.

2.

Monthly Rainfall (in.)	7.6	6.7	8.1	6.2	6.0	6.2

____

3.

Wildcat Basketball Season Wins (number of points won by)							
24	12	10	18	20	12	17	10

____

4.

Tom's Weekly Earnings ($)	200	167	185	212	195	193

____

5. There are seven whole numbers in a data set. The mean of the data set is 28. The median is 29, and the mode is 31. The least number in the data set is 22, and the greatest number is 35. What are the seven numbers in the data set?

____

6. There are seven children in the Arthur family, including one set of twins. The youngest child is 6 years old, and the oldest is 16 years old. The mean of their ages is 11 years, the median is 10 years, and the mode is 15 years. What are the ages of the Arthur children? How old are the twins?

____

Name ______________________ Date __________ Class __________

LESSON 6-2

Reteach

Mean, Median, Mode, and Range

You can find the mean, median, mode, and range to describe a set of data.

Terry's Test Scores	76	81	94	81	78

The **mean** or average is the sum of the items divided by the number of items.

76 + 81 + 94 + 81 + 78 = 410 First, find the sum of the values.
410 ÷ 5 = 82 Then divide the sum by the number of values in the set of data.

The mean is 82 points.

The **median** is the middle value of an ordered set of data. If there are two middle values, the median is the mean of those two values.

76, 78, **81**, 81, 94 Put the values in order first.

The median is 81 points.

The **mode** is the value that occurs most often in a set of data.
The mode is 81 points.

The **range** is the difference between the greatest and least values in the set of data.

94 − 76 = 18 Use subtraction to find the range.

The range is 18 points.

Find the mean, median, mode, and range of each set of values.

1. 23, 78, 45, 22

mean: __________

median: __________

mode: __________

range: __________

2. 102, 79, 82, 103, 79

mean: __________

median: __________

mode: __________

range: __________

3. 56, 99, 112, 112, 56

mean: __________

median: __________

mode: __________

range: __________

Name ______________________ Date ___________ Class ___________

LESSON 6-2

Challenge

Speedy Data

Match each set of data with its description.

Mammal Speeds	
Antelope	54 mi/h
Cheetah	65 mi/h
Greyhound	42 mi/h
Horse	43 mi/h

____

Range Descriptions	
A Range	4 mi/h
B Range	29 mi/h
C Range	23 mi/h
D Range	6 mi/h

Insect Speeds	
Bumble bee	11 mi/h
Honey bee	7 mi/h
Hornet	13 mi/h
Horsefly	9 mi/h

____

Mean Descriptions	
E Mean	49 mi/h
F Mean	42 mi/h
G Mean	10 mi/h
H Mean	51 mi/h

Fish Speeds	
Bluefin tuna	47 mi/h
Bonefish	40 mi/h
Sailfish	69 mi/h
Swordfish	40 mi/h

____

Median Descriptions	
J Median	10 mi/h
K Median	42 mi/h
L Median	43.5 mi/h
M Median	48.5 mi/h

Bird Speeds	
Crane	42 mi/h
Goose	42 mi/h
Mallard	40 mi/h
Swan	44 mi/h

____

Mode Descriptions	
N Mode	none
O Mode	40 mi/h
P Mode	42 mi/h
Q Mode	44 mi/h

Name ______________________ Date ____________ Class ____________

LESSON 6-2

Problem Solving

Mean, Median, Mode, and Range

Write the correct answer.

1. Use the table at right to find the mean, median, mode, and range of the data set.

____

____

____

World Series Winners

Team	Number of Wins
Baltimore Orioles	3
Boston Red Sox	5
Detroit Tigers	4
Minnesota Twins	3
Pittsburgh Pirates	5

2. When you use the data for only 2 of the teams in the table, the mean, median, and mode for the data are the same. Which teams are they?

____

____

Circle the letter of the correct answer.

3. The states that border the Gulf of Mexico are Alabama, Florida, Louisiana, Mississippi, and Texas. What is the mean for the number of letters in those states' names?

A 7 letters

B 7.8 letters

C 8 letters

D 8.7 letters

4. There are 5 whole numbers in a data set. The mean of the data is 10. The median and mode are both 9. The least number in the data set is 7, and the greatest is 14. What are the numbers in the data set?

F 7, 7, 9, 11, and 14

G 7, 7, 9, 9, and 14

H 7, 9, 9, 11, and 14

J 7, 9, 9, 14, and 14

5. If the mean of two numbers is 2.5, what is true about the data?

A Both numbers are greater than 5.

B One of the numbers is less than 2.

C One of the numbers is 2.5.

D The sum of the data is not divisible by 2.

6. Tom wants to find the average height of the students in his class. Which measurement should he find?

F the range

G the mean

H the median

J the mode

Name ______________________ Date __________ Class __________

LESSON 6-2

Reading Strategies

Vocabulary Development

The **mean**, the **median**, the **mode**, and the **range** are measures that describe a set of data.

This picture will help you learn about each measure.

Measures that Describe a Set of Data

Mean (average) Add all the values, then divide by the total number of values.	**Median** The middle value, when the values are arranged in order.
Mode The value that occurs most often in a set of data. There can be one mode, more than one mode, or no mode.	**Range** The difference between the greatest number and the least number in a set of data.

These are Tim's bowling scores from 5 different games:

89 98 110 98 105

Write **mean, median, mode,** or **range** to answer the following questions about Tim's scores.

1. The number 98 indicates which measure?

____

2. Tim added up all his scores and divided by 5. Which measure did he find?

____

3. Tim found that the difference between his highest and lowest score was 21 points. That measure is called the

____

4. Tim noticed that he got a score of 98 twice. Which measure is Tim focusing on?

____

Name ______________________________ Date ____________ Class ____________

LESSON 6-2

Puzzles, Twisters & Teasers

Painting Mystery

For each set of numbers find the mean, the mode, and the median, in that order. Look for the letters in the boxes below and Round each answer to the nearest tenth. To solve the puzzle, read the answers(horizontally) across the mean, mode, and median columns from 1 to 12.

Numbers	Mean	Mode	Median
88, 82, 75, 75, 75	79	75	75
63, 24, 24, 37, 77, 52, 24			
95, 100, 100, 105			
13, 25, 47, 100, 100			
59, 62, 62, 75, 86, 94			
25, 25, 24, 24, 23, 27, 26			
25, 28, 29, 26, 29, 25, 25, 30, 26			
81, 75, 63, 51, 41, 24, 22, 18, 22			
25, 30, 35, 30, 30			
5, 15, 10, 23, 23			
26, 57, 57, 33, 37			
24, 36, 80, 24, 24			

Mean Box
A = 27
M = 43
A = 57
N = 44.1
E = 24.9
O = 79
E = 100
T = 42
L = 15.2
Y = 37.6
L = 73
L = 30

Mode Box
A = 23
N = 75
C = 100
O = 24
D = 22
T = 25
E = 57
T = 24, 25
I = 62
L = 30

Median Box
A = 26
P = 30
E = 75
R = 37
H = 25
R = 100
I = 41
S = 15
K = 47
U = 24
K = 68.5

What did the painter say to the wall?

___ ___ ___ ___ ___ ___ ___ ___ ___ ___ ___ ___

___ ___ ___ ___ ___ ___ ___ ___ ___ ___ ___ ___

___ ___ ___ , ___ ___ ___ ___ ___ ___ ___ ___ ___ ___ ___ !

Name ______________________ Date __________ Class __________

LESSON 6-3

Practice A

Additional Data and Outliers

Use the graph to answer Exercises 1–2.

1. The graph shows how many moons several planets have. Find the mean, median, and mode of the data.

____

____

Planet Moons

Planet	Moons
Earth	●
Mars	● ●
Neptune	● ● ● ● ● ● ● ●
Pluto	●

2. With 18 moons, Saturn has more moons than any other planet in our solar system. Add this number to the data in the graph, and find the mean, median, and mode.

____

____

Use the table to answer Exercises 3–4.

3. The table shows several of the countries that have sent the most astronauts into space. Find the mean, median, and mode of the data.

____

____

Astronauts in Space

Country	Number of Astronauts
France	7
Germany	9
Kazakhstan	2

4. The United States has sent 234 astronauts into space—more than any other country. Add this number to the data in the table and find the mean, median, and mode.

____

____

5. In Exercise 2, which measurement was affected the least by the addition of Saturn's data?

____

6. Is the data in Exercise 4 best described by the mean, median, or mode?

____

Name ______________________ Date ____________ Class ____________

LESSON 6-3

Practice B

Additional Data and Outliers

Use the table to answer Exercises 1–2.

Population Densities

State	People (per mi²)
Idaho	16
Nevada	18
New Mexico	15
North Dakota	9
South Dakota	10

1. The table shows population data for some of the least-crowded states. Find the mean, median, and mode of the data.

____

____

2. Alaska has the lowest population density of any state. Only about 1 person per square mile lives there. Add this number to the data in the table and find the mean, median, and mode.

____

____

Use the table to answer Exercises 3–4.

State Counties

State	Number of Counties
Illinois	102
Iowa	99
North Carolina	100
Tennessee	95
Virginia	95

3. The table shows some of the states with the most counties. Find the mean, median, and mode of the data.

____

____

4. With 254 counties, Texas has more counties than any other state. Add this number to the data in the table and find the mean, median, and mode.

____

____

5. In Exercise 1, which measurement best describes the data? Why is Alaska's population density an outlier for that data set?

____

____

6. In Exercise 4, why is the number of counties in Texas an outlier for the data set? Which measurement best describes the data set with Texas included?

____

____

Name ____________________ Date __________ Class __________

LESSON 6-3

Practice C

Additional Data and Outliers

Use the table to answer Exercises 1–2.

1. The table shows some of the years in which the Super Bowl was won by the most points. Find the mean, median, and mode.

____

____

Super Bowl Winning Margins

Year	Points Won By
1967	25
1972	21
1985	22
1995	23
2001	27

2. The 1990 Super Bowl had the largest winning margin, which was 45 points. Add this number to the data in the table and find the mean, median, and mode. Which best describes the data?

____

____

Use the table to answer Exercises 3–4.

3. The table shows some of the most successful coaches in the NFL. Find the mean, median, and mode.

____

____

Successful NFL Coaches

Coach	Games Won
Paul Brown	170
Bud Grant	168
Chuck Knox	193
Chuck Noll	209
Dan Reeves	179

4. With 347 games won, Don Shula is the most successful NFL coach. Add this number to the data in the table and find the mean, median, and mode. Which best describes the data?

____

____

5. When an outlier is added to a data set, which of these measurements will usually change the most: the range, mean, median, or mode?

____

6. If an outlier is greater than the other data in a set, how will it affect the mean of the data?

____

Name ______________________ Date ____________ Class ____________

LESSON 6-3

Reteach

Additional Data and Outliers

An **outlier** is a value in a set of data that is much greater or much less than the other values.

Number of Minutes Spent on Homework

Mon	Tue	Wed	Thurs	Fri
47	42	45	46	10

The outlier is 10 minutes, because it is much less than the other values in the set.

An outlier may affect the mean, median, or mode.

Data without Friday's value: mean = 45   median = 45.5   no mode
Data with Friday's value: mean = 38   median = 45   no mode

When Friday's value is included, the mean decreases by 7 minutes, the median decreases by 0.5 minutes, and the mode stays the same. The mean is most affected by the outlier because it is less than every value except for the outlier itself.

Find the mean, median, and mode for the set of data with and without the outlier.

1. 22, 25, 48, 26, 21, 27, 26, 29

   With outlier: ______________________

   Without outlier: ______________________

When an outlier affects the mean, median, or mode, choose a value that best describes the data.

In the example above, the median best describes the data because 45 minutes is closer to most of the data values in the set.

Find the mean, median, and mode. Then decide which best describes the set of data.

2. 16, 12, 14, 17, 81, 18, 13, 19, 14, 19

____

____

Name ______________________ Date __________ Class __________

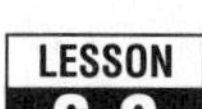

LESSON 6-3

Challenge

Outer Space Outlier

You have been chosen to train as an astronaut! The statisticians at NASA are not happy, because you are a major outlier for their data. Use the information below to find how you will affect their astronaut data.

Youngest Astronauts

In 1970, Russian astronaut Gherman S. Titov became the youngest person to travel into space. He was 25 years old at liftoff. The ages of some of the other youngest astronauts of all time were 26, 29, 28, 27, 26, and 28.

Data Without Your Age:	Data With Your Age:
Mean age:	**Mean age:**
Median age:	**Median age:**
Mode age:	**Mode age:**

Oldest Astronauts

In 1998, American astronaut John H. Glenn became the oldest person to travel into space. He was 77 years old at liftoff. The ages of some of the other oldest astronauts of all time were 54, 59, 61, 56, 58, and 55.

Data Without Your Age:	Data With Your Age:
Mean age:	**Mean age:**
Median age:	**Median age:**
Mode age:	**Mode age:**

Name ______________________ Date __________ Class __________

LESSON 6-3

Problem Solving

Additional Data and Outliers

Use the table to answer the questions.

1. Find the mean, median, and mode of the earnings data.

____

____

Successful Films in the U.S.

Film	U.S. Earnings for first release (million $)
E.T. the Extra-Terrestrial	400
Forrest Gump	330
Independence Day	305
Jurassic Park	357
The Lion King	313

2. *Titanic* earned more money in the United States than any other film—a total of $600 million! Add this figure to the data and find the mean, median, and mode. Round your answer for the mean to the nearest whole million.

____

Circle the letter of the correct answer.

3. In Canada, people watch TV an average of 74 minutes each day. In Germany, people watch an average of 68 minutes a day. In France it is 67 minutes a day, in Spain it is 91 minutes a day, and in Ireland it is 74 minutes a day. Find the mean, median, and mode of the data.

   A mean: 74 min.; median: 74 min.; mode: 74 min.

   B mean: 74 min.; median: 74.8 min.; mode: 74 min.

   C mean: 74.8 min.; median: 74 min.; mode: 24 min.

   D mean: 74.8 min.; median: 74 min.; mode: 74 min.

4. People in the United States watch more television than in any other country. Americans watch an average of 118 minutes a day! Add this number to the data and find the mean, median, and mode.

   F mean: 82 min.; median: 74 min.; mode: 74 min.

   G mean: 82 min.; median: 74 min.; mode: 118 min.

   H mean: 82 min.; median: 91 min.; mode: 74 min.

   J mean: 74.8 min.; median: 82 min.; mode: 74 min.

5. In Exercise 2, which data measurement changed the least with the addition of *Titanic*'s earnings?

   A the range

   B the mean

   C the median

   D the upper extreme

6. In Exercise 4, which measurements best describe the data?

   F mean and median

   G range and mean

   H median and mode

   J range and mode

Name ________________________ Date __________ Class __________

LESSON 6-3

Reading Strategies

Use Graphic Aids

Tim put his bowling scores on a number line.

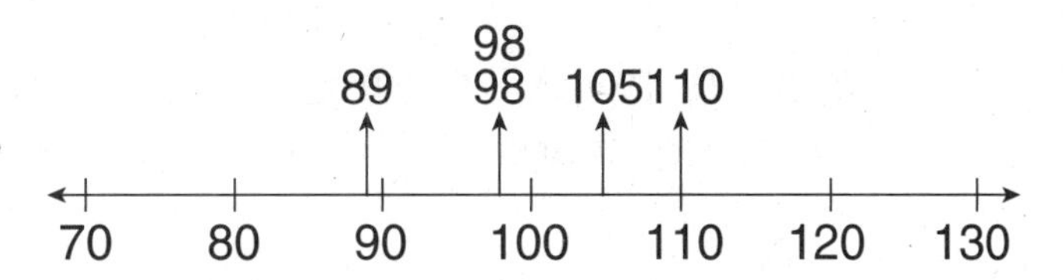

The number line lets you see whether the scores are close together or spread apart.

Recall the measures that describe Tim's scores:

Mean—100 Median—98 Mode—98 Range—21

Tim bowled another game and got a score of 70. This number line includes Tim's new score.

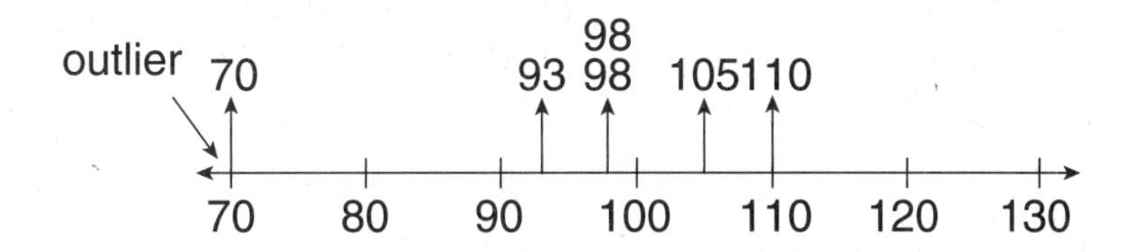

The score of 70 is called an **outlier,** because it is set apart from the other scores.

Answer the following questions.

1. How does the number line help you see the outlier in these scores?

____

2. Will the mean increase or decrease when the score of 70 is included?

____

3. How does the number line help you find the mode?

____

4. With the addition of the score of 70, will the range increase or decrease?

____

5. Circle the correct answer: Which measure is not changed with the added score of 70?

   median mode

Name ______________________________ Date __________ Class __________

Puzzles, Twisters & Teasers

A–Maze–ing Data!

First, answer each question. Then use your answers to navigate through the maze.

1. Consider the data set 2, 4, 6, 8, 8, 8, 10, 12, 14, 16. What happens to the mean if you add the values 3 and 9?

   The mean goes down by _____.

2. What is the mean if you add 2 and 18 to the original data set? _____

3. As CEO of a company you notice that your five executives have the following salaries: $65,000, $70,000, $80,000, $80,000, and $72,000. You are hiring a sixth executive and will pay her $78,000. By how many thousands did the median change? _____

4. Consider the data set 6, 8, 10, 12, 12, 12, 14, 16, 18, 20. What happens to the mean if you add the values 22 and 22.7? The mean goes up by _____.

5. Consider the data set: 6, 8, 12, 13 and 16. Which statistical measure will change if you add 8 and 14 to the data set?

   If Mean move 1 space right
   If Median move 4 spaces left
   If Mode move 5 spaces down
   If none move 3 spaces right

Now you must find your way through this maze to increase the mean of your grades.

Go to start and move down the amount of answer #1 times 10.

Move right the amount of answer #2.

Move up the amount of answer #3.

Move right the amount of answer #4 rounded to the nearest whole number.

Follow the directions for the answer you selected for #5.

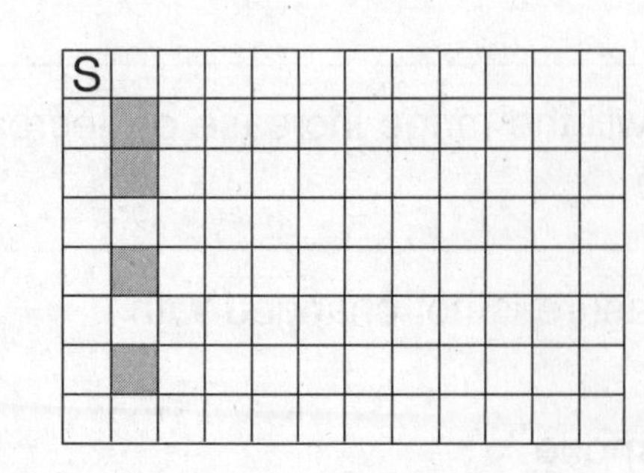

Name ________________________ Date ____________ Class ____________

LESSON **6-4**

Practice A

Bar Graphs

Use the bar graph to answer each question.

1. What is the most common city name in the United States?

   ____

2. Which two names on the graph are used for the same number of cities?

   ____

3. Which name is used for 52 different cities in the United States?

   ____

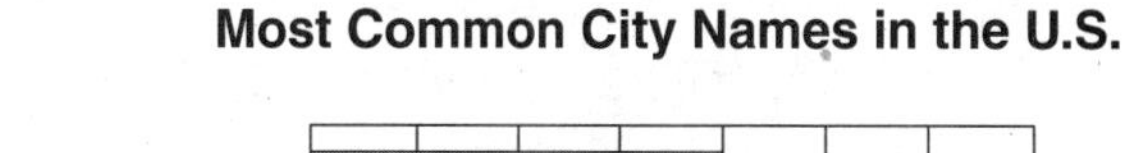

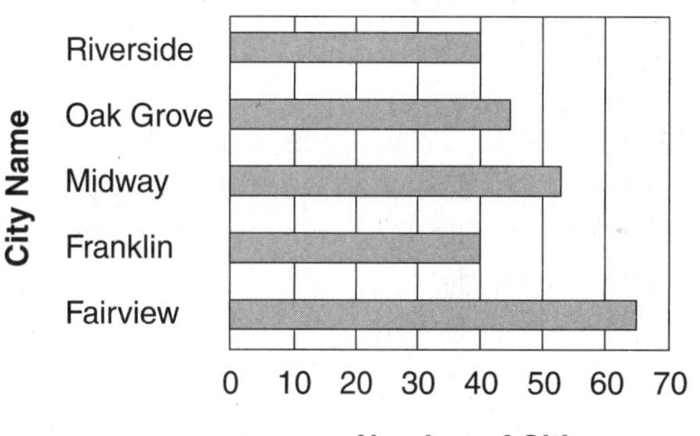

Use the given data to make a bar graph.

Skyscrapers in Some Cities

City, State	Skyscrapers
Chicago, IL	75
Dallas, TX	20
Houston, TX	30
Los Angeles, CA	22
Hong Kong, China	42
Tokyo, Japan	30

Name ______________________ Date ____________ Class ____________

LESSON 6-4

Practice B
Bar Graphs

Use the bar graph to answer each question.

1. In which country did people spend the most money on toys in 2000?

____

2. In which two countries did people spend the same amount of money on toys in 2000? How much did they each spend?

____

____

3. In which country did people spend $9 million on toys in 2000?

____

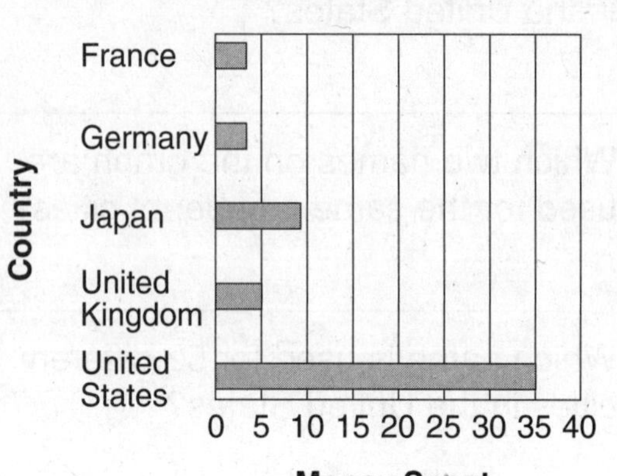

Make a bar graph to compare the data in the table.

Female Groups with the Most Top 10 and Top 20 Hits

Top 10		Top 20	
The Supremes	20	The Supremes	24
The Pointer Sisters	7	The Pointer Sisters	13
TLC	9	TLC	11
En Vogue	5	En Vogue	7
Spice Girls	4	Spice Girls	7

Artist

Key: Top 10 Top 20

Name ______________________ Date ____________ Class ____________

LESSON 6-4

Practice C
Bar Graphs

Use the bar graph to answer each question.

1. How did the number of Asian immigrants coming to the United States compare for 1900 and 2000?

____

2. From which region did most people immigrate in 1900?

____

3. From which region did most people immigrate in 2000?

____

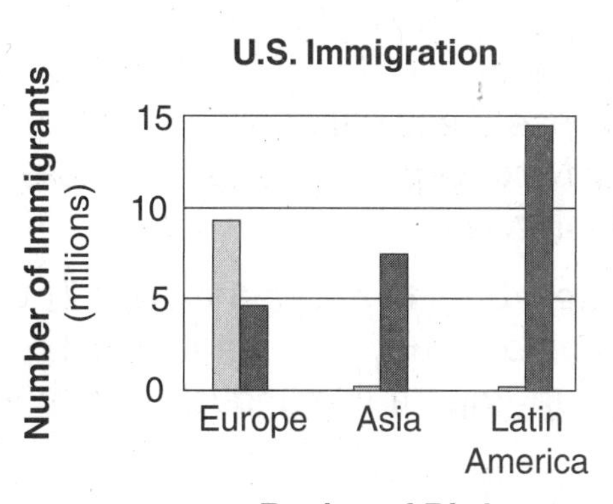

Use the given data to make a bar graph and answer the questions.

U.S. Foreign-born Population

Year	Males	Females
1900	5,630,190	4,711,086
1950	5,258,255	5,089,140
2000	14,200,000	14,179,000

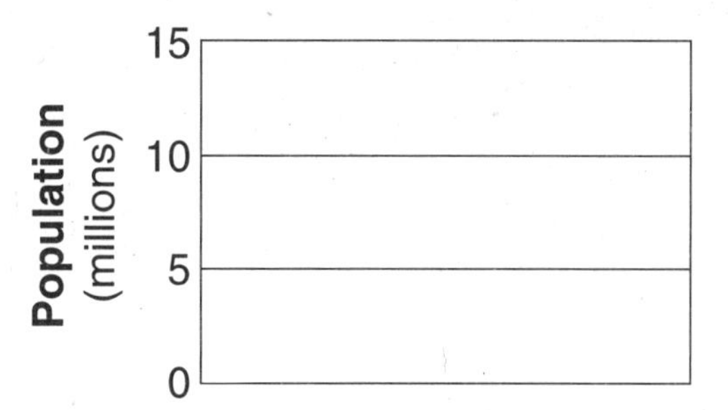

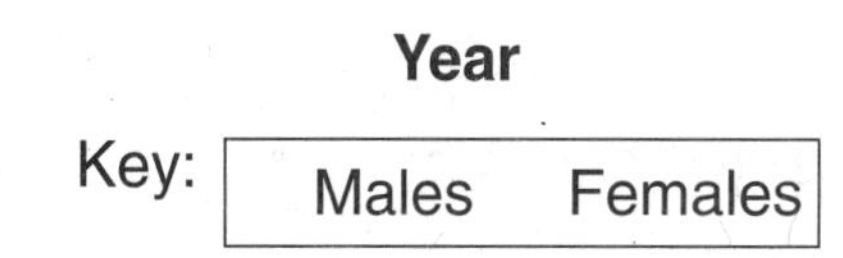

4. Which year had the highest total foreign-born population?

____

5. How did the male and female foreign-born populations compare between 1900 and 2000?

____

____

Name ______________________ Date ____________ Class ____________

LESSON 6-4

Reteach

Bar Graphs

You can make a bar graph to compare amounts.

Annual Read-a-thon Totals			
Grade	6	7	8
Books Read	86	42	98

To make a bar graph using the data in the table, first choose a scale that includes all of the data values. Next, separate the scale into equal parts, called intervals.

Then draw bars to match the data. The bars should be of equal width and should not touch. Give your graph a title and label its axes.

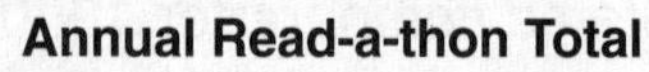

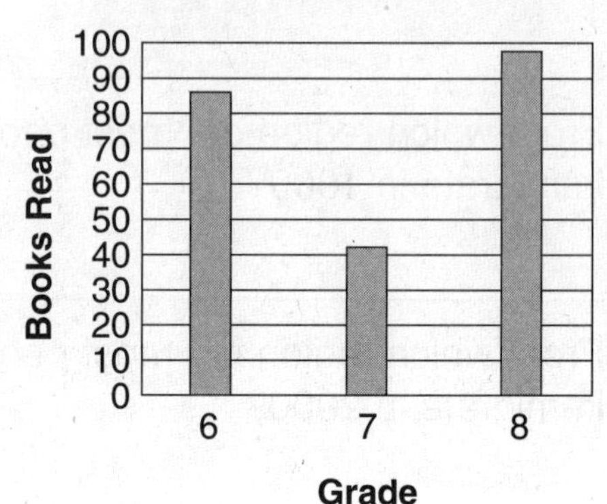

Use the data to make a bar graph.

1.

Canned Food Drive Totals			
Grade	6	7	8
Cans Collected	96	74	62

Look at the bar graph for the Read-a-thon above. Which grade read almost twice as many books as the seventh grade?

The bar for the sixth grade is about twice a long as the bar for the seventh grade. So the sixth grade read almost twice as many books as the seventh grade.

Use the bar graph you made in Exercise 1.

2. How many more items did the sixth grade collect than the eighth grade?

____

Name ______________________ Date __________ Class __________

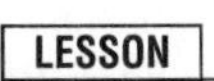

Reteach

Bar Graphs (continued)

A double-bar graph shows two sets of related data.

To make a double-bar graph, choose a scale and an interval for the scale. Then draw bars to match the data. The bars for the same grade should touch, but bars for different grades should not touch.

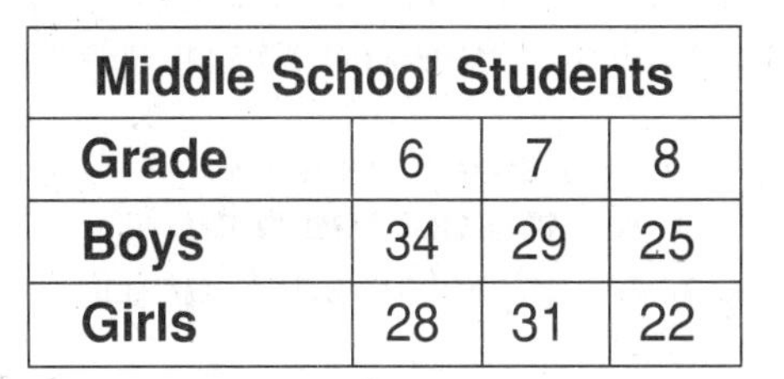

Middle School Students			
Grade	6	7	8
Boys	34	29	25
Girls	28	31	22

Because there are two bars for each grade, make a key to show which bars represent girls and which bars represent boys.

Your graph should have a title and its axes should be labeled.

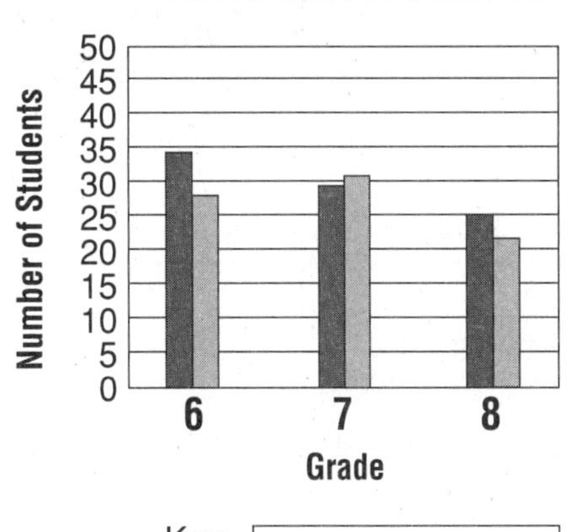

Make a double-bar graph to match the data below. Then answer the question.

3.

Movie Tickets Sold			
	Fri	**Sat**	**Sun**
Adult	136	118	98
Student	84	102	154

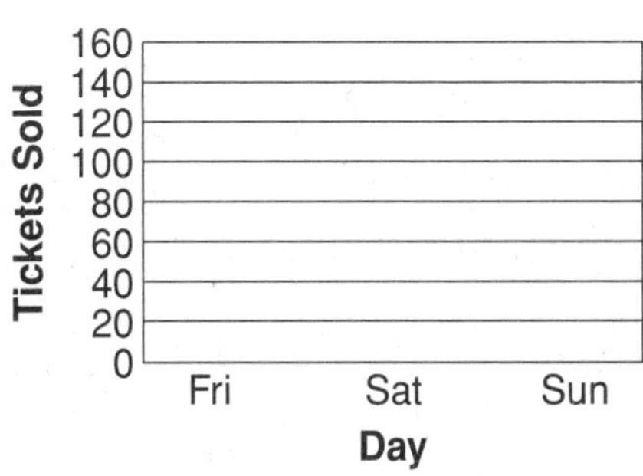

Key: ____________

4. On which day was the number of adult tickets sold about the same as the number of student tickets sold?

____

____

Name ______________________ Date ____________ Class ____________

LESSON 6-4

Challenge

Picture a Bar Graph

Sometimes people use illustrations instead of bars to display data on bar graphs. For example, a bar graph showing the sizes of some forests might use trees for the graph's bars.

Use the data given in each table below to make an illustrated bar graph. Make sure the pictures you choose for your bars relate to the subject of each graph.

Fastest U.S. Roller Coasters

Roller Coaster, Location	**Speed** (mi/h)
Desperado, NV	80
Goliath, CA	85
Millennium Force, OH	92
Superman: The Escape, CA	100
Titan, TX	85

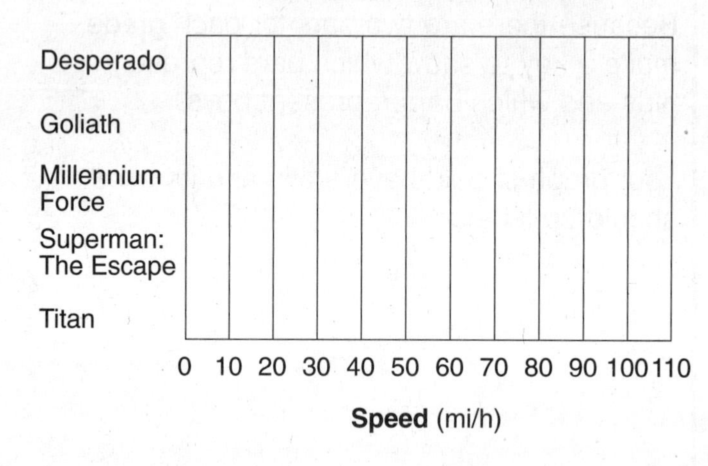

Surfers with the Most World Championship Wins

Surfer, Country	**Wins**
Tom Carroll, Australia	2
Tom Curren, United States	3
Damien Hardman, Australia	2
Mark Richards, Australia	4
Kelly Slater, United States	6

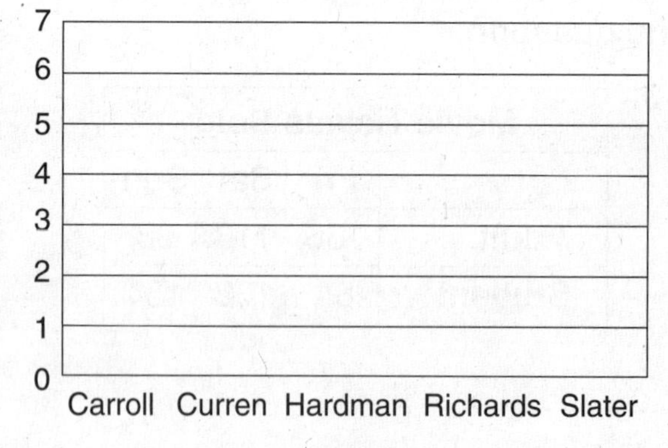

Name ______________________ Date __________ Class __________

LESSON 6-4

Problem Solving

Bar Graphs

Use the bar graph for Exercises 1–4.

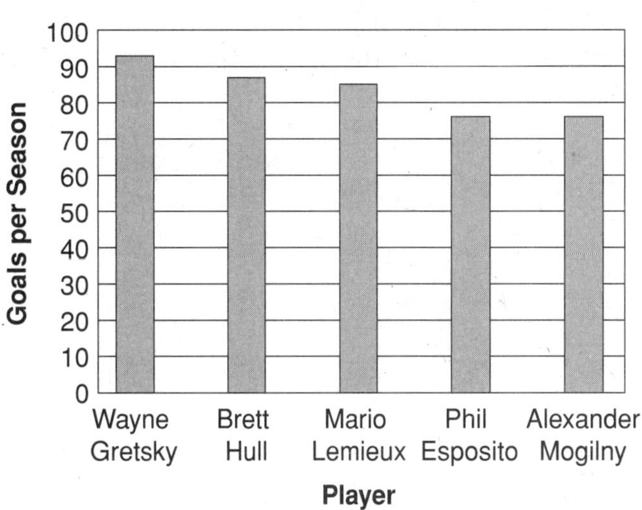

1. What is the range of the goals the hockey players scored per season?

   ____

2. What is the mode of the goals scored?

   ____

3. What is the mean number of goals the players scored?

   ____

Use the bar graph for Exercises 5–8.

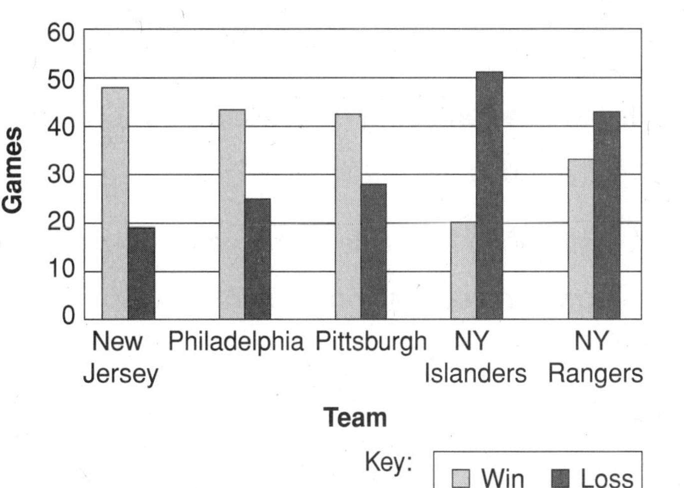

4. Which team won the most games that season? ______________

5. Which team lost the most games that season? ______________

6. What was the mean number of games won? ______________

7. What was the mean number of games lost? ______________

Circle the letter of the correct answer.

8. Which hockey team had the greatest difference between the number of games won and lost?

   A New Jersey
   B New York Islanders
   C Philadelphia
   D Pittsburgh

9. How do you know the mode of a data set by looking at a bar graph?

   F The mode depends on the type of data displayed in the bar graph.
   G The mode is the first bar.
   H The mode has the lowest bar.
   J The bar for the mode is in the middle of the graph.

Name ______________________ Date ____________ Class ____________

LESSON 6-4

Reading Strategies

Compare and Contrast

A **bar graph** shows data and makes it possible to compare facts about that data. A group of sixth-grade students voted on their favorite kind of exercise. The graph shows the data that was collected from the votes.

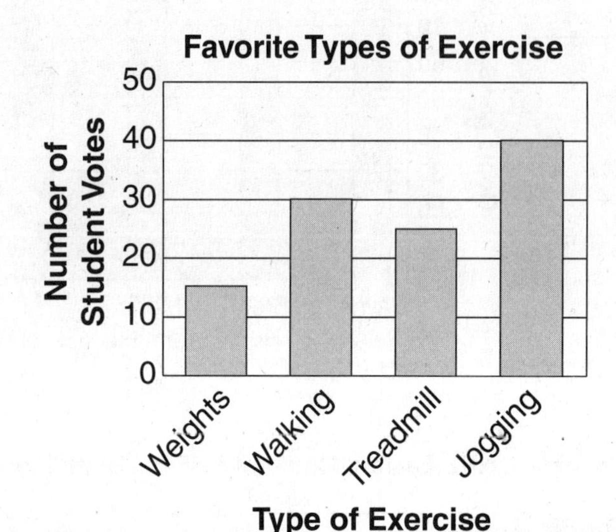

Answer the following questions about the bar graph.

1. What does the graph show? ______________________

2. What does the scale of the graph count by? ______________________

3. How can you tell from the graph which exercise had the most votes?

____

4. Compare votes for walking and votes for jogging. Which one got the most votes?

5. Compare votes for the treadmill and votes for weights. Which one had the most votes? ______________________

6. Which exercise had the least amount of votes on the graph? ______________________

7. How can you tell from looking at the graph which exercise had the fewest votes?

____

8. How do the bars on the graph help you compare data?

____

Name ______________________ Date ___________ Class ___________

LESSON **6-4**

Puzzles, Twisters & Teasers

At The Movies

Ten students were surveyed to see how many times they went to the movies last month.

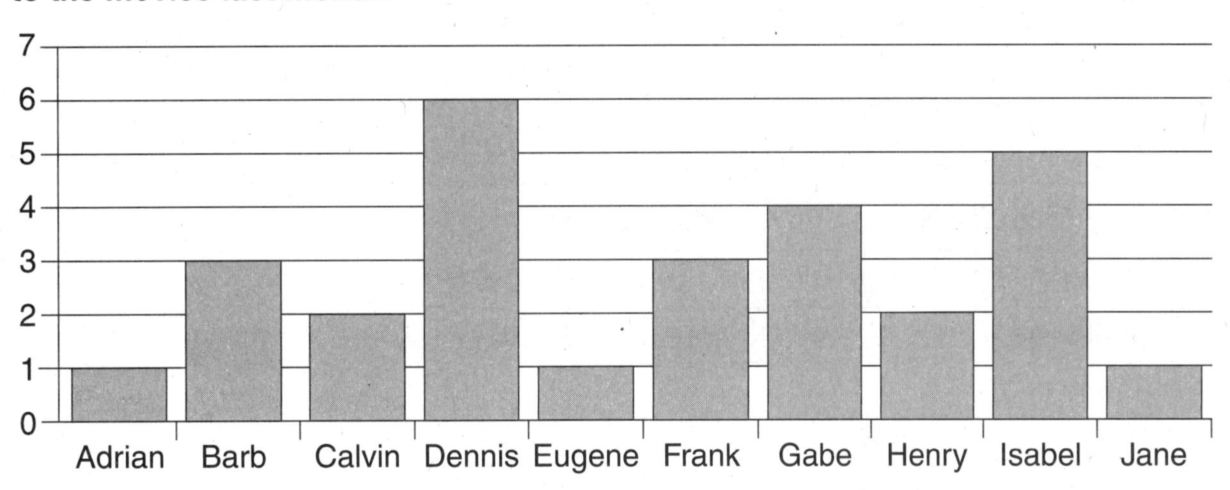

To solve the riddle answer each question. Use the letter in front of each answer to fill in the blanks of the riddle.

1. How many students went to the movies more than four times last month?

S = 1 **T** = 2 **L** = 3 **R** = 4

2. Who went to the movies the most last month?

I = Calvin **E** = Gabe **A** = Dennis **O** = Isabel

3. How many students went to the movies less than three times last month?

R = 3 **S** = 4 **B** = 5 **C** = 6

4. How many students went to the movies only once last month?

W = 0 **L** = 1 **K** = 2 **U** = 3

5. How many more times did Dennis go to the movies than Gabe?

K = 2 **N** = 3 **G** = 4 **D** = 5

6. Frank and Barb went to the movies a total of how many times?

L = 3 **T** = 4 **A** = 5 **C** = 6

7. Who went to twice as many movies as Calvin?

Y = Barb **N** = Frank **O** = Gabe **D** = Isabel

Where do Australian children go to play?

___ ___ ___ ___ ___ ___ ___
7 4 1 3 2 6 5

Name ______________________ Date ___________ Class ___________

LESSON 6-5

Practice A

Line Plots, Frequency Tables, and Histograms

1. Hockey players voted for a team name. The results are shown in the box. Make a tally table. Which name got the fewest votes?

____

Bears	Wildcats	Bulldogs	Lions	Bears	Wildcats
Bears	Bears	Wildcats	Bears	Lions	Bears

Tally Table for Hockey Team Name Votes

Bears	
Bulldogs	
Lions	
Wildcats	

2. Make a line plot of the data.

Number of Goals Scored by 25 Hockey Players											
0	2	4	2	1	0	2	5	3	2	1	3
0	1	4	3	1	2	3	1	4	5	1	2

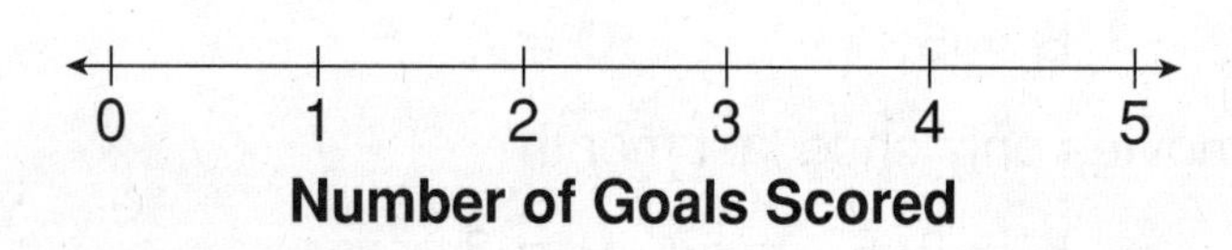

3. Use the data in the box below to complete the frequency table with intervals.

Ages of Hockey Fans Polled at Tonight's Game									
14	10	38	54	27	29	7	16	10	45
18	21	9	36	25	17	39	33	26	30

Ages of Hockey Fans Polled at Tonight's Game						
Ages	1–10	11–20	21–30	31–40	41–50	51–60
Frequency						

4. To which age group did the most fans belong? ______________

Name ______________________ Date __________ Class __________

LESSON 6-5

Practice B

Line Plots, Frequency Tables, and Histograms

1. Students voted for a day not to have homework. The results are shown in the box. Make a tally table. Which day got the most votes?

____

Monday	Friday	Thursday	Friday	Tuesday	Friday
Friday	Thursday	Wednesday	Monday	Friday	Monday

Tally Table for Homework Votes

Mon	
Tues	
Wed	
Thurs	
Fri	

2. Make a line plot of the data.

Average Time Spent on Homework Per Day (min)									
20	21	24	20	21	20	20	22	25	20
22	20	24	25	24	25	25	21	25	24

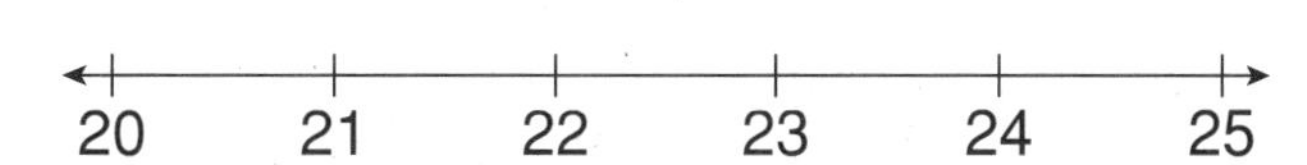

Average Time Spent on Homework Per Day (min)

3. Use the data in the box below to make a frequency table with intervals.

Class Social Studies Test Scores									
78	95	81	83	75	68	100	73	92	85
59	70	88	92	99	87	75	67	89	84

Class Social Studies Test Scores					
Scores					
Frequency					

4. In which range of scores did most of the students' tests fall? ______________

Name ______________________ Date ____________ Class ____________

LESSON 6-5

Practice C

Line Plots, Frequency Tables, and Histograms

1. During a car trip, Ed counted the number of red, black, blue, and green cars he passed on the highway. The results are shown in the box. Make a tally table. Which color of car did Ed count most often?

____

red	black	blue	green	red	blue	red	black
blue	green	red	black	red	red	black	

Tally Table for Car Colors

Red	
Black	
Blue	
Green	

2. Make a line plot of the data.

Ages of Drivers at Truck Stop										
17	28	29	34	47	55	27	18	39	42	22
33	21	46	52	55	41	48	19	23	25	

16 17 18 19 20 21 22 23 24 25 26 27 28 29 30 31 32 33 34 35 36 37 38 39 40 41 42 43 44 45 46 47 48 49 50 51 52 53 54 55

Ages of Drivers at a Truck Stop

3. Use the data in the box above to make a frequency table with intervals.
4. Use your frequency table from Exercise 3 to make a histogram.
5. To which age group did most of the drivers belong? __________

Ages of Drivers at Truck Stop

Age Intervals	Frequency

Ages of Drivers

Name ______________________ Date __________ Class __________

LESSON 6-5

Reteach

Line Plots, Frequency Tables, and Histograms

Julie picked the following cards from a deck.

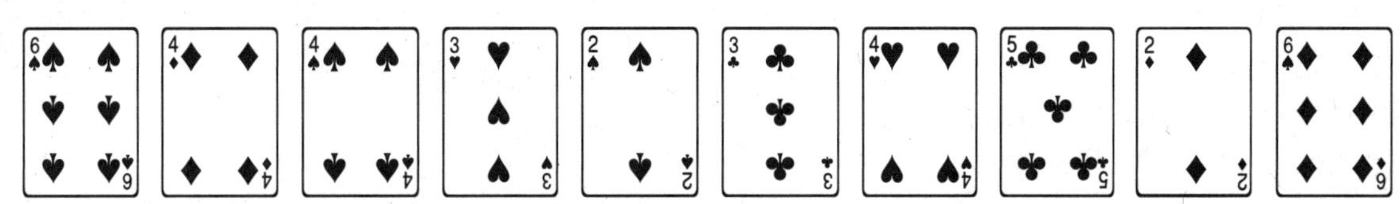

You can make a tally table to organize the data. Make a row for the numbers. Then for each card, make a tally mark in the appropriate column.

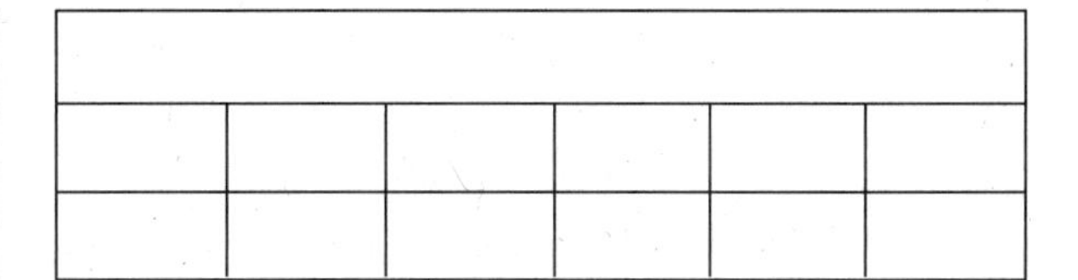

Julie's Cards

2	3	4	5	6
||	||	|||	|	||

1. Make a tally table to organize the data.

Rolls of a Number Cube						
2	3	6	5	1	4	1
3	3	5	1	6	1	4

A line plot gives a visual picture of data. To make a line plot of Julie's data, draw a number line. Then use an X to represent each tally mark in the tally table.

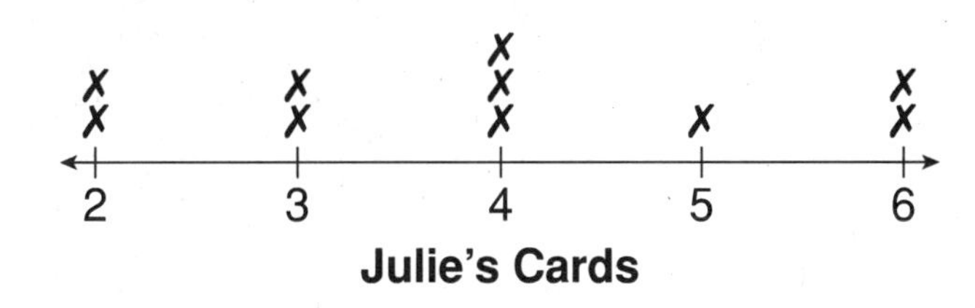

2. Make a line plot for the tally table you made Exercise 1.

Name ______________________ Date ____________ Class ____________

LESSON 6-5

Reteach

Line Plots, Frequency Tables, and Histograms

Sometimes, you can make a frequency table with intervals or a histogram.

Number of Jumping Jacks Completed in 30 Seconds				
12	28	24	32	35
31	38	55	43	52
42	49	18	22	15
47	37	19	31	37

A frequency table can organize the data with intervals.

Jumping Jacks

Interval	Frequency
1–10	0
11–20	4
21–30	3
31–40	7
41–50	4
51–60	2

A histogram is a bar graph that shows the number of values that occur within each interval.

You make a histogram the same way you make any other bar graph, except that the bars touch. They do not overlap.

Here is a histogram for the frequency table above.

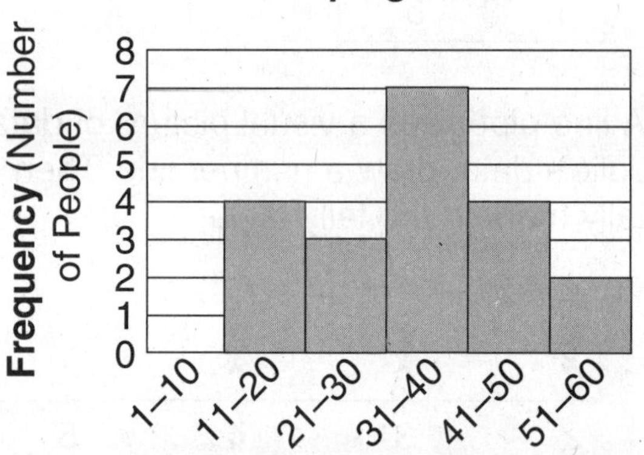

3. Use the data to make a histogram.

Total Books Read by Participants in Summer Reading Program				
5	3	8	7	6
2	9	10	1	2
4	5	7	3	5
3	1	0	10	4
3	5	8	2	1
1	7	0	4	11

12
10
8
6
4
2
0

Name ______________________________ Date ___________ Class ___________

LESSON **6-5**

Challenge

Write Often

What letter is used more than any other letter in the English language? ______

The box below contains the six English letters that are used most often. Use the box to complete the tally table at the bottom of the page. Your completed table will show the answer to the question.

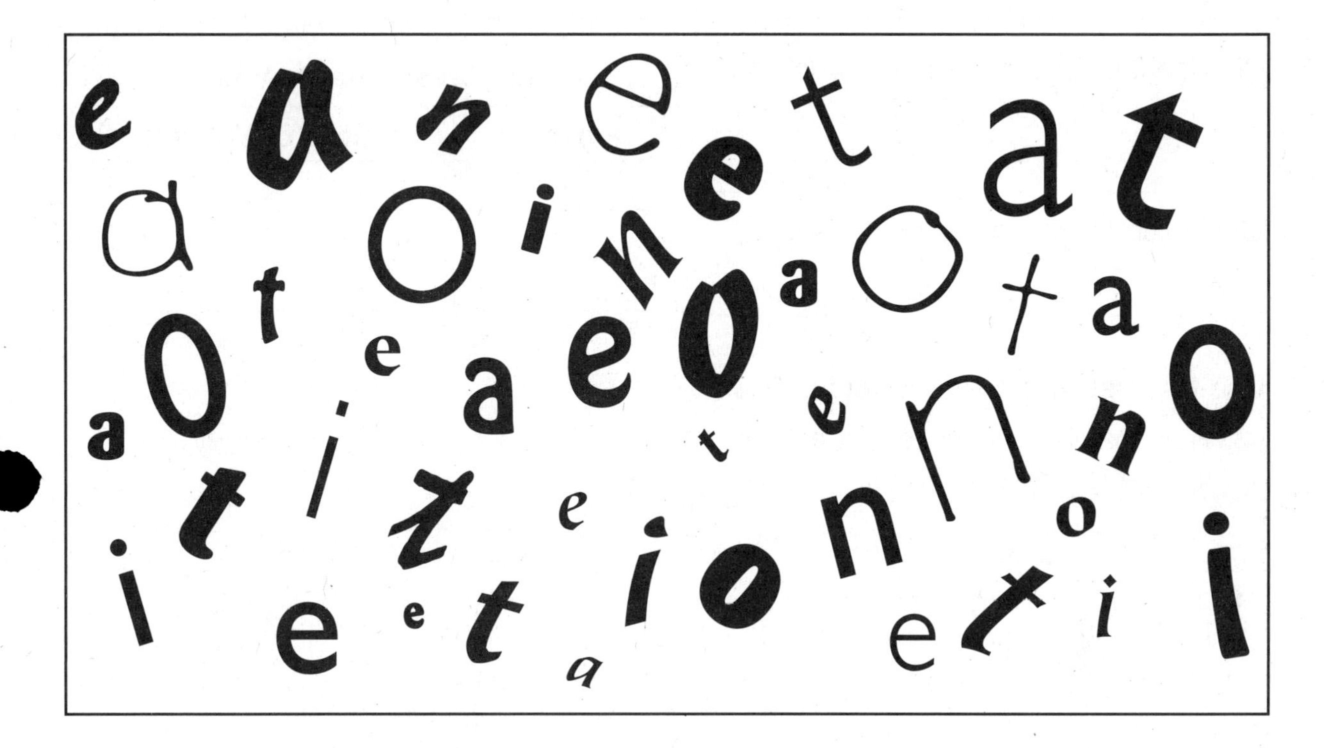

Tally Table for the Number of Letters	
A	
E	
I	
N	
O	
T	

Name ______________________ Date ____________ Class ____________

LESSON 6-5

Problem Solving

Line Plots, Frequency Tables, and Histograms

The sixth grade class voted on their favorite ice cream flavors. The results of the vote are shown below.

chocolate	vanilla	strawberry	vanilla	vanilla
vanilla	chocolate	vanilla	chocolate	strawberry
chocolate	strawberry	vanilla	vanilla	chocolate

1. Use the data to make a tally table. How many students voted in all?

____

2. Which flavor got the most votes?

____

Ice Cream Flavor Votes

Flavor	Number of Votes

Use the histogram for Exercises 3–5.

3. How many years make up each age interval on the histogram?

____

4. Which range of ages on the histogram has the highest population?

____

5. Which range of ages has the lowest population?

____

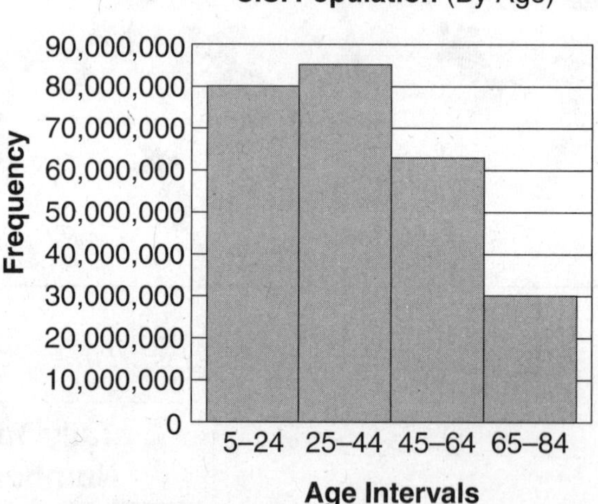

Circle the letter of the correct answer.

6. Which of the following cannot be used to make a frequency table with intervals?
   A histogram
   B tally table
   C line plot
   D double-bar graph

7. Which question can be answered by using the histogram above?
   A How many people in the United States are younger than 5 years?
   B What is the mean age of all people in the United States?
   C How many people in the United States are older than 84 years old?
   D How many people in the United States are age 25 to 64?

Name ______________________ Date ____________ Class ____________

LESSON **6-5**

Reading Strategies

Reading a Table

This is a **tally table** showing the number of cartons of milk sold to sixth-graders in one week.

Day	Number of Milk Cartons
Monday	𝍸 𝍸 𝍸 II
Tuesday	𝍸 𝍸 𝍸
Wednesday	𝍸 𝍸 II
Thursday	𝍸 𝍸 III
Friday	𝍸 𝍸 IIII

Answer each question about the tally table.

1. How are the tally marks organized? ______________________
2. For how many days are milk sales shown? ______________________
3. How can you tell by looking at the table which day had the most sales?

____

____

Creating a **frequency table** is one way to summarize the data in the tally chart. This table organizes data by how often it occurs.

Day	**Frequency** (Number of cartons sold)
Monday	17
Tuesday	15
Wednesday	12
Thursday	13
Friday	14

Use the frequency table to answer the following questions.

4. How is the data in the tally table shown differently than data in the frequency table?

____

5. How many cartons of milk were sold on Tuesday? ______________________
6. After lunch on Wednesday, how many cartons of milk had been sold so far that week? ______________________

Name ______________________ Date __________ Class __________

LESSON 6-5

Puzzles, Twisters & Teasers

Bertha's Burger Bonanza

Bertha's menu includes hamburgers, cheeseburgers, hot dogs, chilidogs, French fries, milk shakes, chicken, juice, and tacos. One day, to speed up things, Bertha herself wrote down everything the people in line wanted that day. Here is her list.

Chili Dog	Juice	Tacos	Cheeseburger	Chicken
Hamburger	Juice	Cheeseburger	Juice	Hamburger
Hot Dog	French Fries	Juice	Hot Dog	Chicken
Milk Shake	Chili Dog	Juice	Juice	Chicken
French Fries	Hamburger	Tacos	French Fries	Chili Dog
Hamburger	Chicken	Cheeseburger	Cheeseburger	Hamburger
Tacos	Hamburger	Chicken	Tacos	Hamburger
French Fries	Chicken	French Fries	Tacos	Juice
Hamburger	Tacos	Tacos	Tacos	Tacos

Complete the table below using the data from Bertha's list.

Number of Foods Ordered at Bertha's

Type of Food	Frequency
Hamburger	
Cheeseburger	
Hot Dog	
Chili Dog	
French Fries	
Chicken	
Milk Shake	
Tacos	
Juice	

To solve the riddle, find the letter that corresponds to each frequency. Rearrange the letters to solve the riddle. ** The letter for chicken is used twice. **

8	18	1	32	7	9	19	5	16	23	6	30	4	7	13	2	28	22	11	15	3	18
U	D	E	K	R	P	B	N	W	G	I	Z	T	M	F	A	S	C	J	Y	H	O

The skydiver had fries and a milk shake. Why was he nervous?

Because he was ___ ___ ___ ___ ___ ___ ___ ___ ___ ___!

Name ______________________ Date ___________ Class ___________

LESSON 6-6

Practice A

Ordered Pairs

Use the coordinate grid to answer Exercises 1–2. Circle the letter of the correct answer.

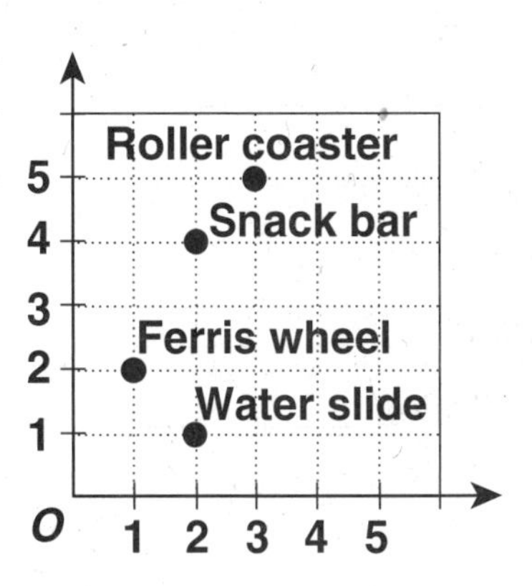

1. Which ordered pair gives the location of the roller coaster in the park?
   A (5, 3)
   B (3, 3)
   C (3, 5)
   D (5, 5)

2. Which feature on the coordinate grid of the park is located at point (1, 2)?
   F Ferris wheel
   G Roller coaster
   H Snack bar
   J Water slide

Use the coordinate grid to answer Exercises 3–4. Circle the letter of the correct answer.

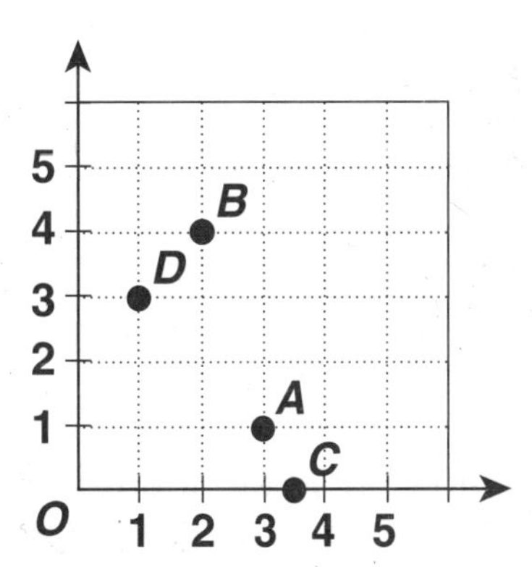

3. Which ordered pair describes the location of point *C*?
   A (1, 3)
   B (3, 1)
   C $(3\frac{1}{2}, 0)$
   D (0, 3)

4. Which point is located at (1, 3) on the grid?
   F point A
   G point B
   H point C
   J point *D*

5. The restrooms at the amusement park are located at point (4, 4). Label the restrooms on the grid for Exercises 1–2.

6. Point *E* is located 1 unit up and 2 units to the right of point *A* on the grid above. Label point *E* on the grid for Exercises 3–4.

Name ______________________ Date ____________ Class ____________

LESSON 6-6

Practice B

Ordered Pairs

Name the ordered pair for each location on the grid.

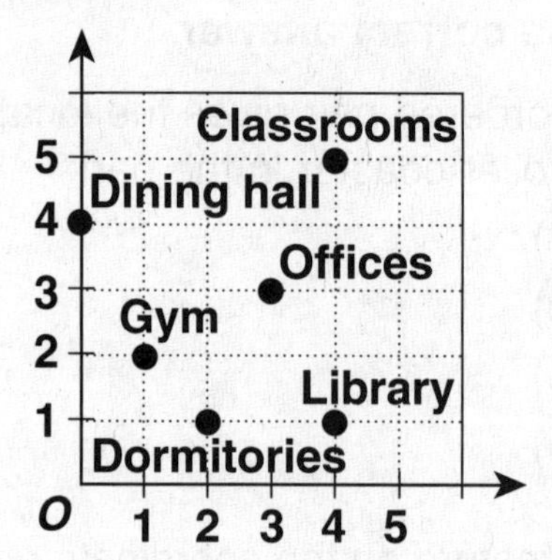

1. gym ____________
2. dining hall ____________
3. offices ____________
4. library ____________
5. classrooms ____________
6. dormitories ____________

Graph and label each point on the coordinate grid.

7. $A\ (5, 1\frac{1}{2})$
8. B (2, 2)
9. C (1, 3)
10. D (4, 3)
11. E (5, 5)
12. F (2, 4)

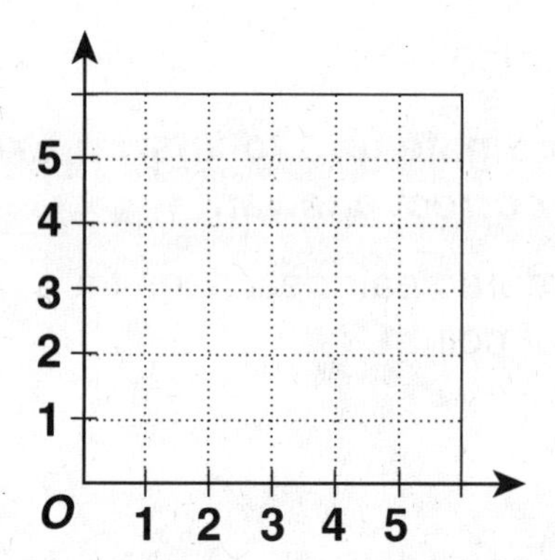

13. On a map of his neighborhood, Mark's house is located at point (7, 3). His best friend, Cheryl, lives 2 units west and 1 unit south of him. What ordered pair describes the location of Cheryl's house on their neighborhood map?

____

14. Quan used a coordinate grid map of the zoo during his visit. Starting at (0, 0), he walked 3 units up and 4 units to the right to reach the tiger cages. Then he walked 1 unit down and 1 unit left to see the pandas. Describe the directions Quan should walk to get back to his starting point.

____

____

Name ______________________ Date ____________ Class ____________

LESSON 6-6

Practice C

Ordered Pairs

Give the ordered pair for each location.

1. firehouse ____________

2. police station ____________

3. city hall ____________

4. zoo ____________

5. park ____________

6. post office ____________

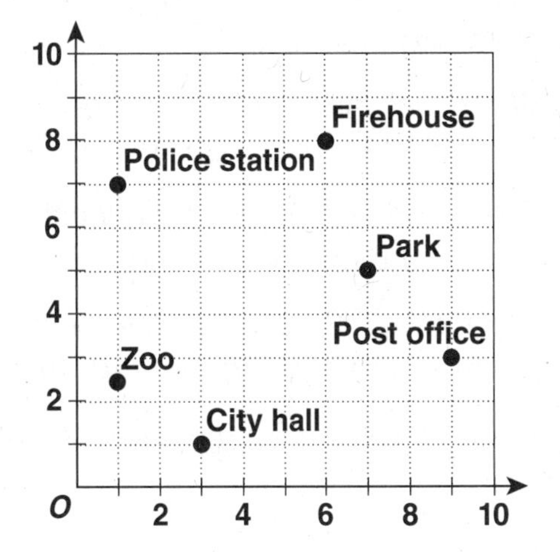

Name the point found at each location.

7. (6, 2) ______

8. (9, 2) ______

9. (9, 5) ______

10. (6, 5) ______

11. (1, 6) ______

12. $(3, 8\frac{1}{2})$ ______

13. (5, 6) ______

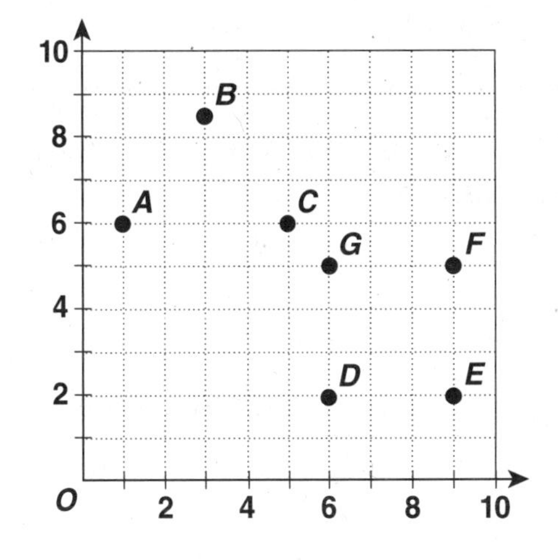

14. Draw lines to connect the points *A* and *B*, *B* and *C*, and *C* and *A*. What shape is formed?

____

15. Draw lines to connect the points *D* and *E*, *E* and *F*, *F* and *G*, and *G* and *D*. What shape is formed?

____

16. If you wanted to create a square using points *A* and *C* on the grid above, how many new points would you have to draw? What ordered pairs describe those new points?

____

Name ______________________ Date __________ Class __________

LESSON 6-6

Reteach

Ordered Pairs

A coordinate grid is formed by horizontal and vertical lines and is used to locate points.

An ordered pair names the location of a point by using two numbers.

The ordered pair (2, 5) gives the location of point *A* on the coordinate grid.

The first number, 2, tells the horizontal distance from the starting point (0, 0).

The second number, 5, tells the vertical distance.

To find the ordered pair for point *B*, start at (0, 0). Then move 6 units right and $3\frac{1}{2}$ units up. The coordinates of point *B* are $(6, 3\frac{1}{2})$.

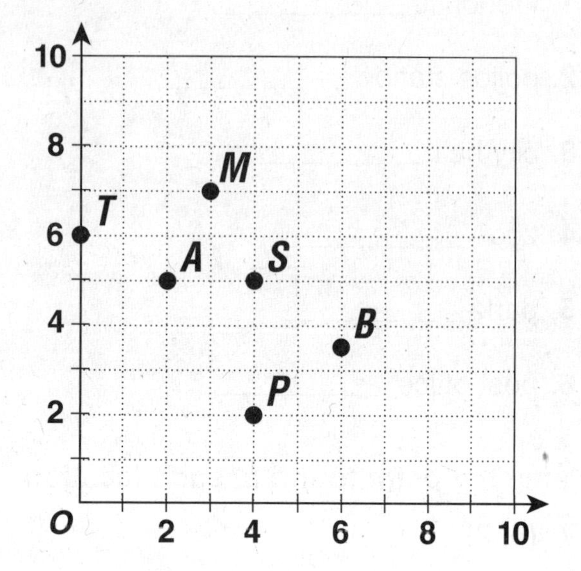

Give the ordered pair for each point shown on the coordinate grid above.

1. *P* ____________

2. *T* ____________

3. *M* ____________

4. *S* ____________

You can plot points in a coordinate grid.

To plot *F* (6, 4), start at (0, 0). Then move 6 units right and 4 units up.

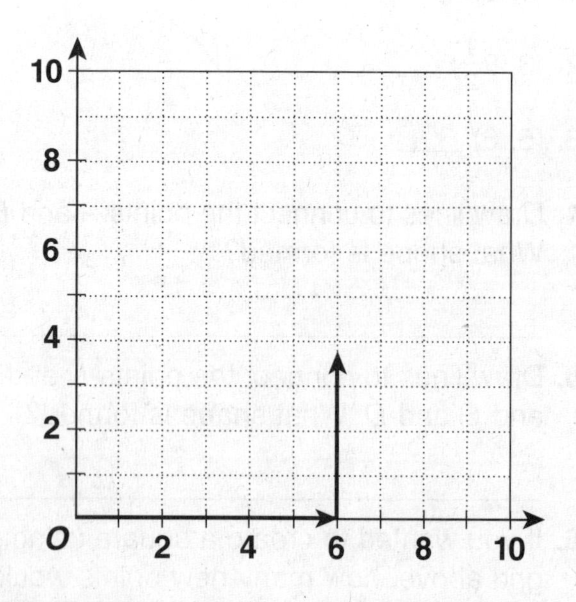

Plot each point in the coordinate grid above.

5. *V* (5, 6)

6. *G* (3, 2)

7. *K* (7, 0)

8. $C(1, 5\frac{1}{2})$

Name ______________________ Date __________ Class __________

LESSON 6-6

Challenge

Treasure Island

According to legend, a pirate named Blackbeard buried his stolen treasures somewhere on the Outer Banks, off the coast of North Carolina. While on vacation there, you found an old sea chest buried on the beach. Inside it was a map that just may lead you to Blackbeard's hidden treasures!

Follow the map's clues to each location on the island. For each clue, name the location and the ordered pair that describes its point on the map. Graph each of those points on the map. Then draw an *X* where you think the treasure is buried.

1. Land your ship 7 units east and 6 units north of the (0, 0) starting point. ____________
2. Walk 5 units west. ____________
3. Walk 5 units south and 1 unit east. ____________
4. Walk 2 units east and 2 units north. ____________
5. Walk 2 units north and 1 unit east. ____________
6. Walk 1 unit east and 3 units north. ____________
7. Walk 3 units west and 1 unit north to find my treasure. ____________

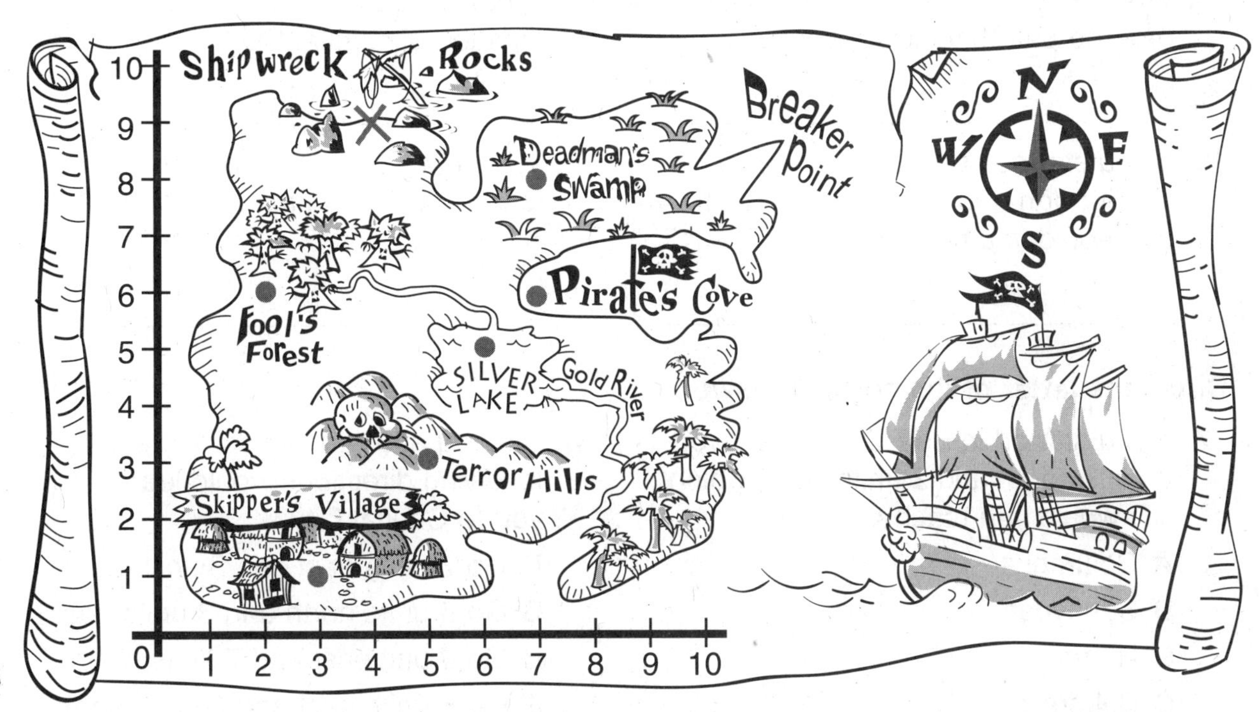

Name ______________________ Date ____________ Class ____________

LESSON 6-6

Problem Solving

Ordered Pairs

Use the coordinate grid to answer each question.

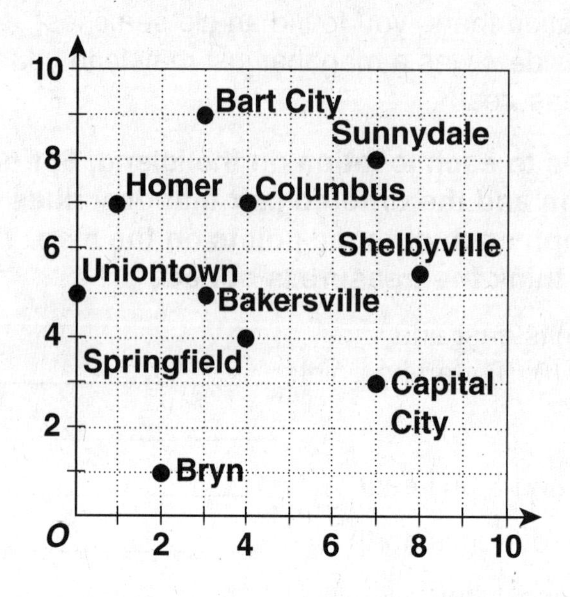

1. What city is located at point (4, 4) on the map?

____

2. Which city is located at point $(8, 5\frac{1}{2})$ on the map?

____

3. Which city's location is given by an ordered pair that includes a 0?

____

4. What ordered pair describes the location of Capital City?

____

5. If you started at (0, 0) and moved 1 unit north and 2 units east, which city would you reach?

____

6. Which two cities on the map are both located 4 units to the right of (0, 0)?

____

Circle the letter of the correct answer.

7. If you started in Bart City and moved 2 units south and 2 units west, which city would you reach?

A Columbus

B Sunnydale

C Homer

D Bakersville

8. Starting at (0, 0), which of the following directions would lead you to Capital City?

F Go 7 units east and 3 units north.

G Go 5 units north and 3 units east.

H Go 3 unit east and 7 units north.

J Go 8 units east and 6 units north.

Name ______________________________ Date ___________ Class ___________

LESSON 6-6

Reading Strategies
Sequencing Directions

A **coordinate grid** is formed by a series of horizontal and vertical lines. Each point on the grid can be shown by using an **ordered pair.**

Reading a map of the city is like finding a point on a **coordinate grid.**

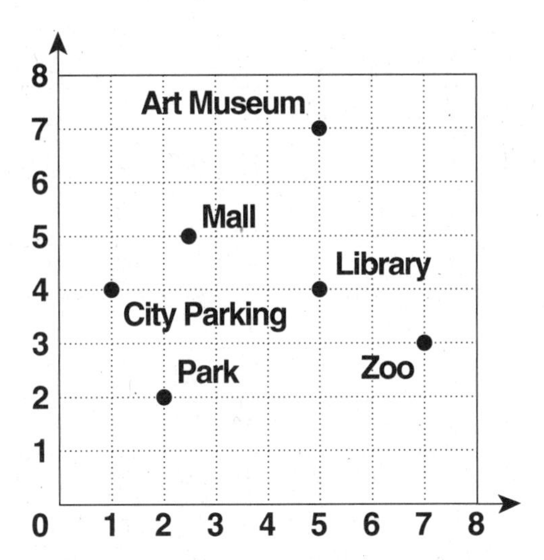

This sequence of steps helps you find points on the coordinate grid:
Step 1: Start at 0 on the coordinate grid.
Step 2: The first number tells you how far to move to the right.
Step 3: The second number tells you how far to move up.
For example, the coordinates (7, 3) mean:

Start at 0. Move right 7 units. Go up 3 units. ⟶ You're at the zoo!

Answer each question.

1. Write the sequence of steps that will get you to the library.

____

2. What ordered pair gives the location of the library?

3. Write the sequence of steps that will get you to the mall.

____

4. What ordered pair gives the location of the mall? ____________________

5. The word *sequence* means to put things in order. Why do you think the numbers used to graph a point on the grid are called ordered pairs?

____

____

Name ______________________ Date ____________ Class ____________

LESSON 6-6

Puzzles, Twisters & Teasers

Griddle

In a griddle (grid riddle) the answer to the riddle is hidden within the grid. To solve the riddle, locate the letters in the grid by the given ordered pairs.

18	G	Y	A	S	D	F	N	A	V	K	L	V	Y	B	L	R	A	V
17	H	U	Q	T	E	Z	A	O	H	T	O	R	V	R	O	U	B	O
16	F	F	E	A	T	B	U	D	S	K	U	K	M	A	C	F	F	R
15	I	A	W	S	I	C	E	R	E	I	L	C	R	T	Y	E	C	X
14	D	U	M	I	U	J	Y	T	Y	M	T	I	T	B	E	N	O	O
13	U	A	G	T	Y	S	R	A	I	B	U	T	I	A	I	N	L	T
12	A	F	R	S	D	W	E	L	E	C	O	E	M	E	S	T	O	Z
11	H	X	E	H	A	B	T	O	R	K	J	W	Y	S	E	O	I	M
10	C	E	K	R	U	M	A	T	H	H	A	E	M	M	A	T	I	C
9	D	I	D	H	R	D	O	N	U	P	T	S	A	R	E	O	E	S
8	U	E	H	E	O	A	T	G	C	Z	N	W	R	R	B	E	E	R
7	J	F	M	E	U	N	W	R	O	T	O	A	C	L	F	I	Z	U
6	R	U	A	B	L	T	E	E	B	L	R	T	U	R	R	F	I	T
5	Z	T	A	T	K	D	L	E	E	X	E	N	S	O	U	E	N	D
4	O	N	U	O	E	O	S	O	C	F	A	F	Q	A	B	P	E	Z
3	Q	G	A	F	U	O	T	F	A	N	D	I	M	W	I	L	D	R
2	H	R	P	T	R	E	R	N	F	P	N	P	L	N	X	K	L	N
1	U	O	V	O	T	N	R	E	K	V	A	R	L	W	E	L	M	T
	1	2	3	4	5	6	7	8	9	10	11	12	13	14	15	16	17	18

Why did Antonio quit his job at the doughnut store?

Because he decided

___ ___   ___ ___ ___ ___
(2, 5) (4, 1)   (13, 4) (9, 9) (12, 3) (18, 1)

___ ___ ___   ___ ___ ___ ___
(2, 5) (3, 8) (7, 6)   (1, 2) (4, 1) (11, 15) (5, 4)

___ ___ ___ ___ ___ ___ ___ ___.
(10, 13) (9, 9) (7, 4) (12, 3) (17, 5) (5, 4) (15, 12) (4, 15)

Name ______________________ Date __________ Class __________

LESSON **6-7**

Practice A

Line Graphs

Use the line graph to answer each question.

1. In which month shown on the line graph does Washington usually receive the most precipitation?

____

2. In general, how does precipitation in Washington, D.C., change between August and October?

____

3. In which months does the city usually receive the same amount of precipitation?

____

Washington, D.C., Average Normal Precipitation

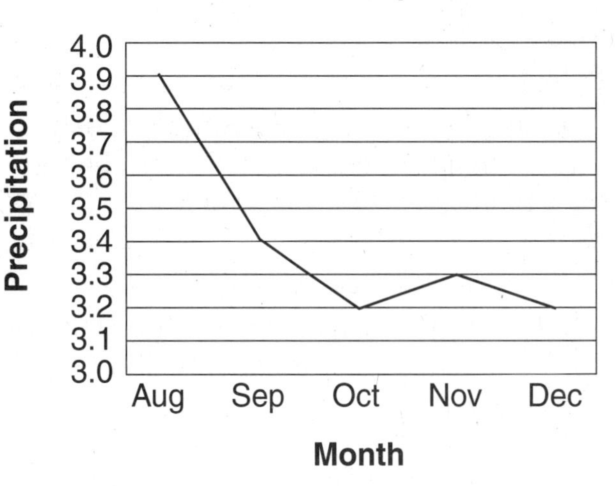

4. Use the given data to make a line graph.

Washington, D.C., Average Normal Temperatures

Month	Temperature (°F)
January	31
March	43
May	62
July	76
September	67
November	45

Washington, D.C., Average Normal Temperatures

Month

Name ______________________ Date __________ Class __________

LESSON 6-7

Practice B
Line Graphs

Use the line graph to answer each question.

1. In which year were the average weekly earnings in the United States the highest?

____

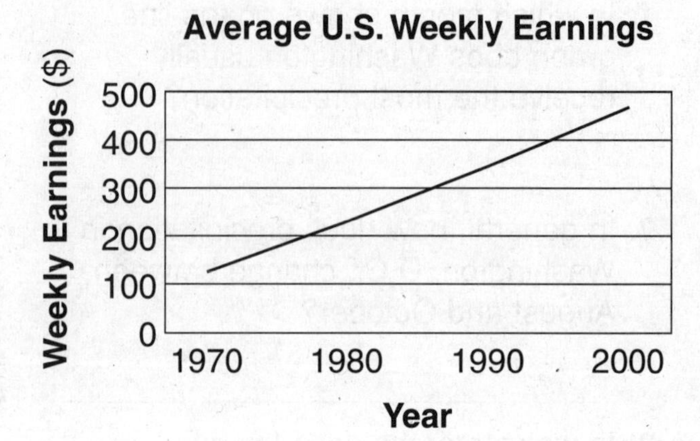

2. In general, how did average weekly earnings in the United States change between 1970 and 2000?

____

3. In which year did the average United States worker earn about $350 a week?

____

4. Use the given data to make a line graph.

U.S. Minimum Wage

Year	Hourly Rate
1970	$1.60
1980	$3.10
1990	$3.80
2000	$5.15

5. Between which two years shown on the graph did the U.S. minimum wage change the least?

____

6. How has the hourly minimum wage changed in the U.S. since 1970?

____

Name ______________________ Date __________ Class __________

LESSON 6-7

Practice C
Line Graphs

Use the double-line graph to answer each question.

1. In which year were the populations of Baltimore and Philadelphia the closest?

____

2. Since 1950, what has happened to the populations of both cities?

____

3. In which year did the cities have their highest populations?

____

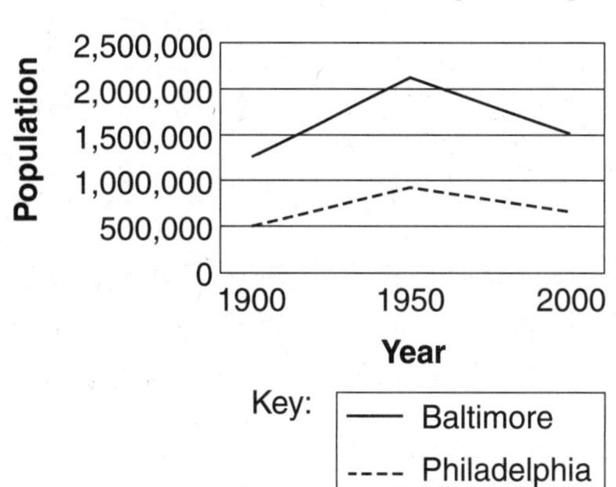

4. Use the given data to make a double-line graph.

Maryland and Pennsylvania Populations

Year	Maryland	Pennsylvania
1900	1,188,044	6,302,115
1950	2,343,001	10,498,012
2000	5,296,486	12,281,054

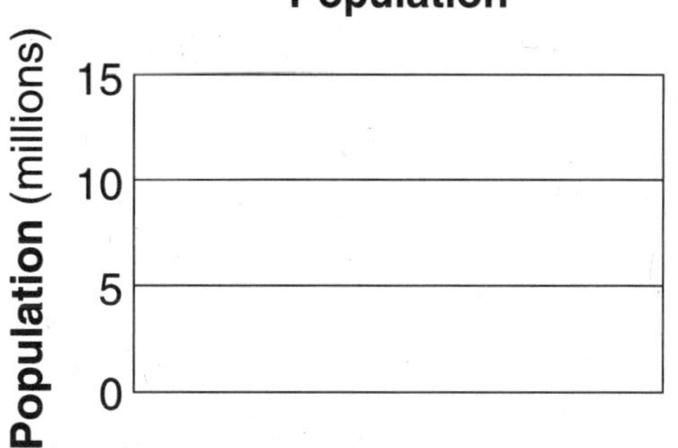

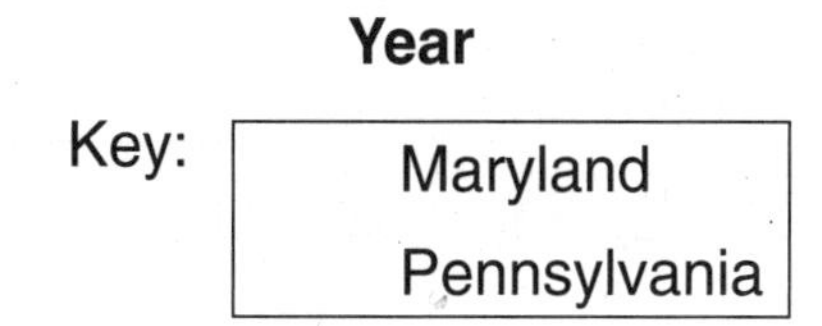

5. In which year were the populations of Maryland and Pennsylvania the closest?

____

6. How did the populations of the two states compare between 1900 and 2000?

____

____

Name ______________________ Date __________ Class __________

LESSON 6-7

Reteach

Line Graphs

A line graph shows change over time.

You can represent data by making a line graph.

Stock Sales (millions)

Mon	Tue	Wed	Thurs	Fri
1.5	2.0	2.25	1.75	0.5

To make a line graph, make "days" the horizontal axis and "sales" the vertical axis. Label the axes.

Then determine an appropriate scale and interval for each axis.

Think of the data in the table as ordered pairs. Mark a point for each ordered pair. Then connect the points with straight segments.

Make sure your line graph has a title.

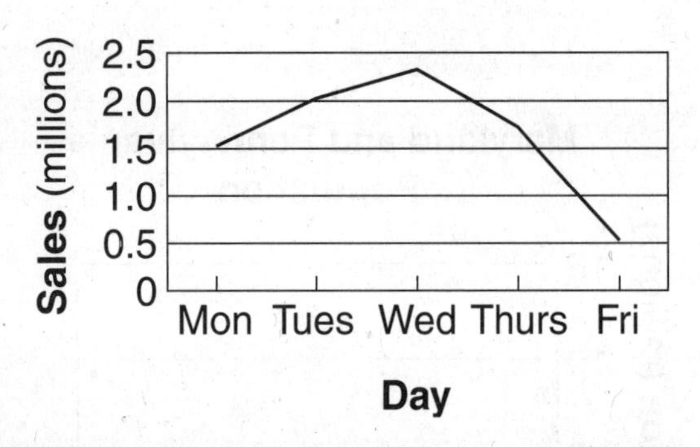

Use the data in the table to make a line graph.

1. **Millie's Savings Account**

Jan	Feb	March	April	May
30	40	35	45	25

Name ______________________________ Date ___________ Class ___________

Reteach

Line Graphs (continued)

Sometimes, you need to make a double-line graph to represent data.

Stock Sales (millions)

Stock	Jan	Feb	March	April	May
A	1.5	2	2.25	1.75	0.5
B	1	2.5	2	1.5	0.75

To make a double-line graph, follow the same steps for making a line graph. Mark and connect points for each of the two sets of data you are displaying. Because there are two sets of data, make a key. Be sure to title the graph and label the axes.

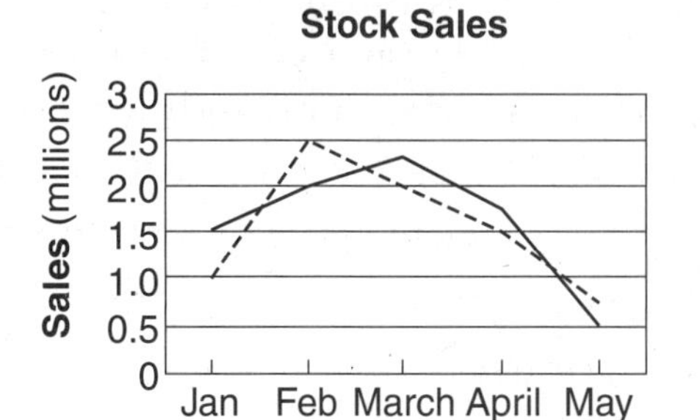

Key:

——Stock A
----Stock B

Use the data in the table to make a double-line graph.

2. **Savings Account**

Student	Jan	Feb	March	April	May
Michael	20	10	30	25	35
Janet	30	20	35	25	45

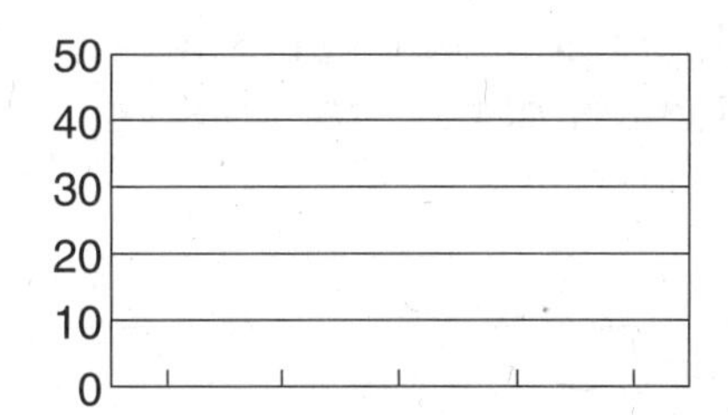

Key:

——Michael
----Janet

Name ______________________ Date ____________ Class ____________

LESSON **6-7**

Challenge

A Trendy Park

Because line graphs show changes over time, you can use them to make predictions based on trends, or patterns. United States park rangers make line graphs to look for trends. They count the number of people who visit their parks each month. Then the park rangers analyze the data on line graphs to look for trends and predict how many visitors to expect each month in the coming years. This data helps the rangers schedule workers and provide services for their visitors.

Great Smoky Mountains National Park in Tennessee and North Carolina receives more visitors each year than any other national park. Imagine you are a park ranger there. Use the line graph below to identify trends and make predictions about the number of visitors the park will receive in the future.

Visitors At Great Smoky Mountains National Park, 2000

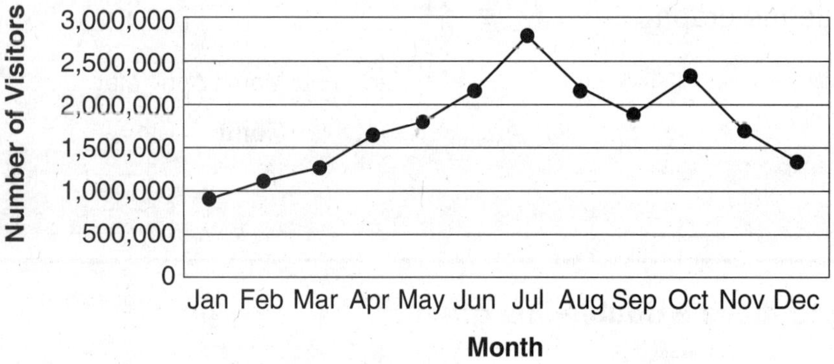

Number of Visitors

3,000,000
2,500,000
2,000,000
1,500,000
1,000,000
500,000
0

Jan Feb Mar Apr May Jun Jul Aug Sep Oct Nov Dec

Month

1. In which month next year should you plan for the most visitors at your park? ____________

2. What can you expect next year at the park between October and January?

____

3. You are in charge of deciding how many rangers should be scheduled to work at Great Smoky Mountains National Park each month next year. How will the number of park rangers you schedule change each month from January to June?

____

____

4. Which month next year would be best for you to take time off from your park ranger job and go on your own vacation? Explain.

____

____

Name ______________________ Date __________ Class __________

LESSON 6-7

Problem Solving

Line Graphs

Use the line graphs to answer each question.

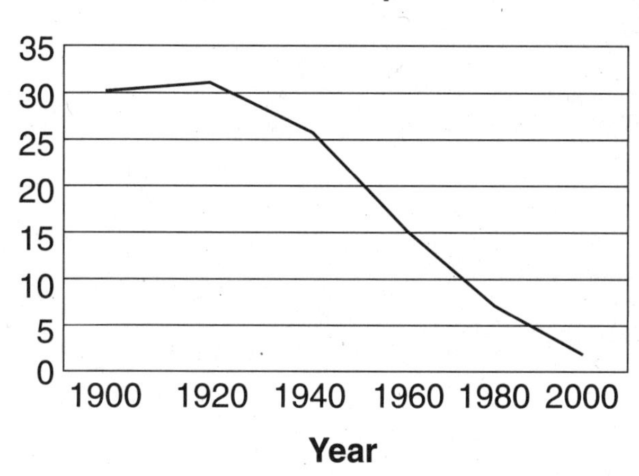

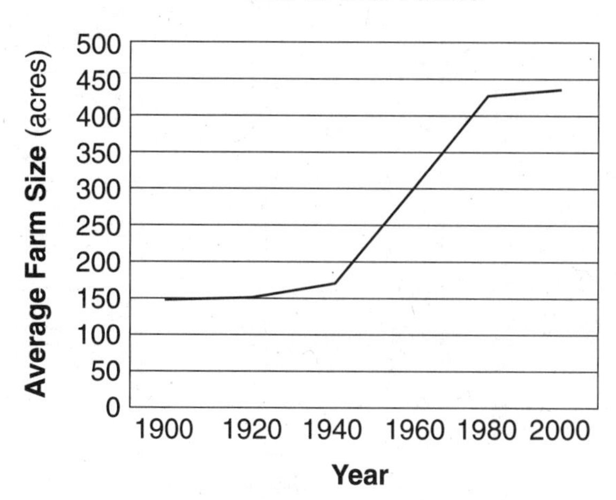

1. In which year was the U.S. farm population the highest? the lowest?

____

2. In which year was the size of the average U.S. farm the largest? the smallest?

____

3. In general, how has the U.S. farm population changed in the last 100 years?

____

4. In general, how has the size of the average U.S. farm changed in the last 100 years?

____

Circle the letter of the correct answer.

5. How many people lived on farms in the United States in 1940?
   - **A** 31 million
   - **B** 30 million
   - **C** 26 million
   - **D** 15 million

6. How many acres did the average farm in the United States cover in 1980?
   - **F** 150 acres
   - **G** 300 acres
   - **H** 400 acres
   - **J** 426 acres

7. Between which two years did the U.S. farm population increase?
   - **A** 1900 and 1920
   - **B** 1920 and 1940
   - **C** 1940 and 1960
   - **D** 1960 and 1980

8. Between which two years did the average size of farms in the United States change the least?
   - **F** 1900 and 1920
   - **G** 1920 and 1940
   - **H** 1960 and 1980
   - **J** 1980 and 2000

Name ______________________ Date ____________ Class ____________

LESSON 6-7

Reading Strategies

Reading a Graph

A **line graph** shows how data changes over a period of time. The line graph below shows a family's weekly food costs for four weeks.

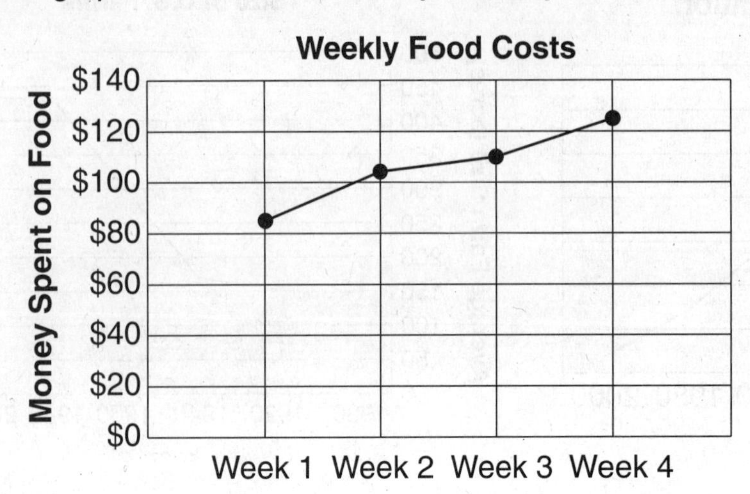

Answer the following questions about the line graph.

1. What information is located along the left side of the graph?

____

2. By what amount do the dollars increase on the left side of the graph?

____

3. What information is shown along the bottom of the graph?

____

Each point on the graph identifies the amount of money spent, by week.

4. How much money was spent on food in Week 1?

____

5. How much money was spent on food in Week 4?

____

6. From the line graph, what can you conclude about food costs for this family?

____

Name ______________________ Date __________ Class __________

LESSON 6-7

Puzzles, Twisters & Teasers

Frogs, Frogs, Frogs!

Every summer, from 1994 to 2004, three brothers would go to the lake and catch frogs. The graph shows the record they kept of their catches.

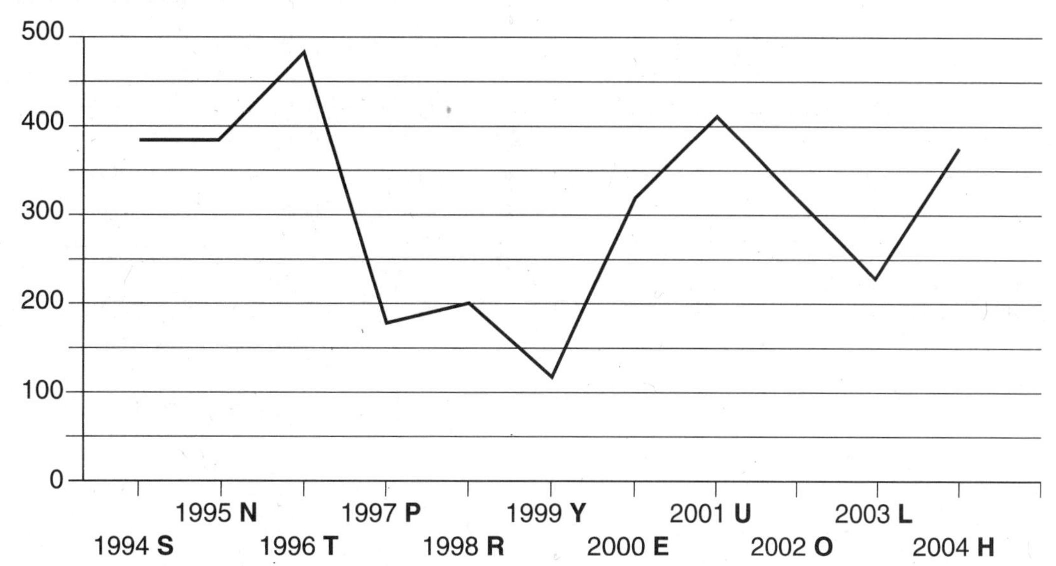

Find the year that answers the questions below and the letter beside that year. Once you have the answer, put the letter that is next to the answer over the question number in the riddle answer.

1. The year they caught about 175 frogs. ______________
2. The only year when they caught the same number of frogs as the previous year. ______________
3. The year they caught the least number of frogs. ______________
4. The only year when they caught about 150 more frogs than the year before. ______________
5. The year when they had the biggest difference from the year before. ______________
6. The year they caught the second most frogs. ______________
7. The year when they caught less than the year before but more than the year after. ______________

How did the frog feel when he broke his leg?

___ ___ ___ ___ ___ ___ ___
6 2 4 7 5 1 3

Name ______________________ Date ____________ Class ____________

LESSON 6-8

Practice A

Misleading Graphs

Use the graph to answer each question.

1. Why is this bar graph misleading?

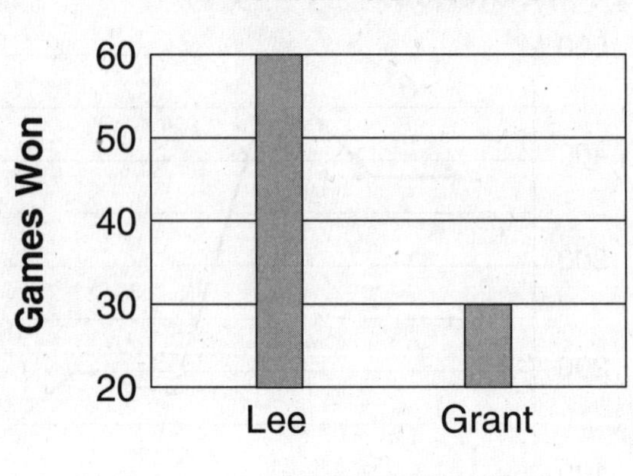

2. What might people believe from the misleading graph?

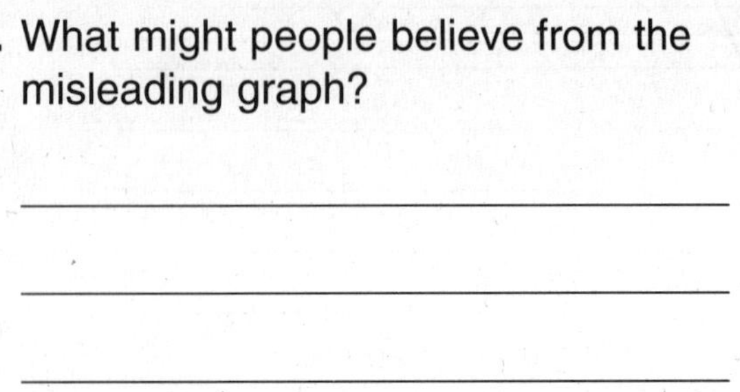

Use the graph to answer each question.

3. Why is this line graph misleading?

____

____

____

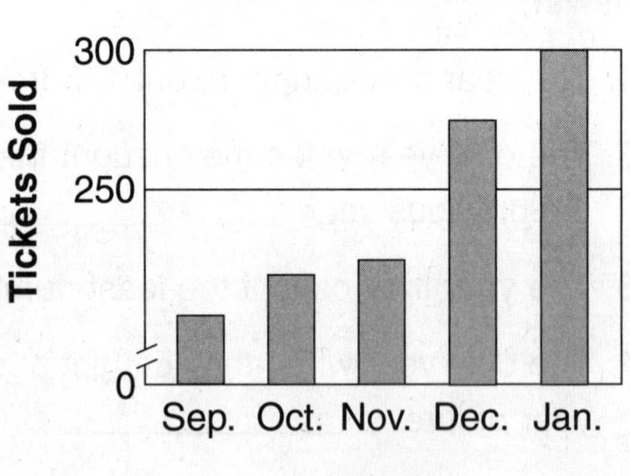

4. What might people believe from the misleading graph?

____

____

____

Name ______________________ Date ____________ Class ____________

LESSON **6-8**

Practice B
Misleading Graphs

Use the graph to answer each question.

1. Why is this bar graph misleading?

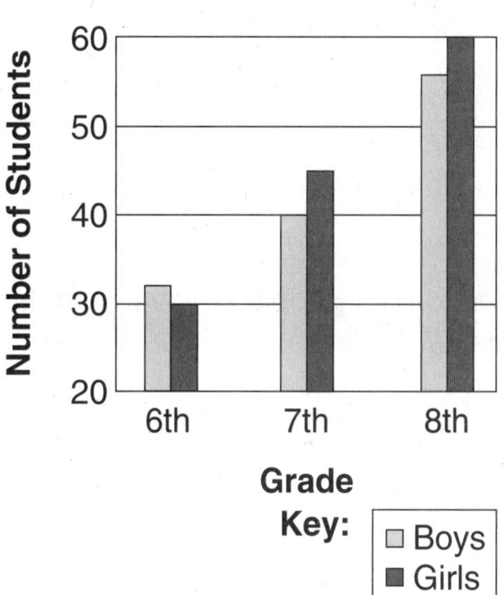

____

____

____

____

2. What might people believe from the misleading graph?

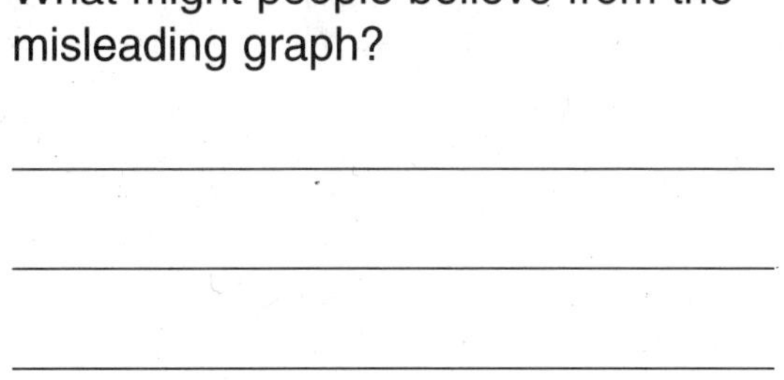

____

____

____

Use the graph to answer each question.

3. Why is this line graph misleading?

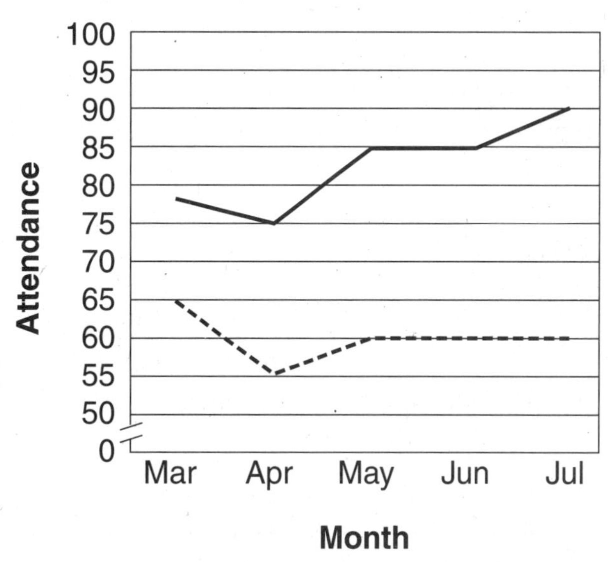

____

____

____

____

4. What might people believe from the misleading graph?

____

____

____

Name ________________ Date __________ Class __________

LESSON 6-8

Practice C

Misleading Graphs

Use the graph to answer each question.

1. Why is this bar graph misleading?

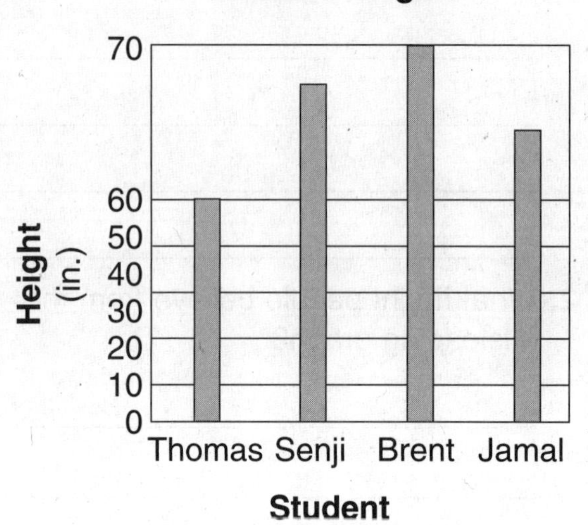

2. What might people believe from the misleading graph?

Use the graphs to answer each question.

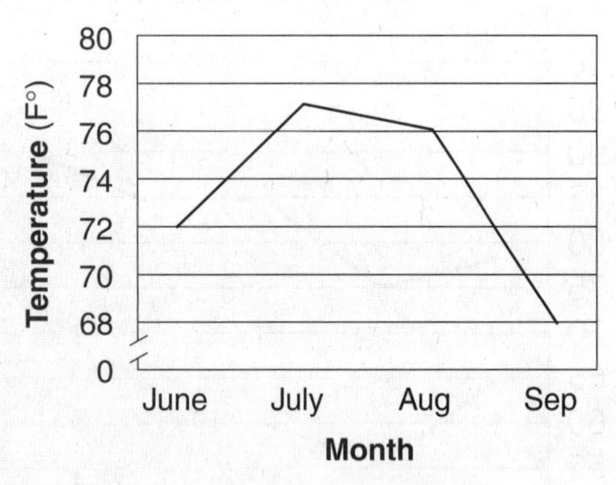

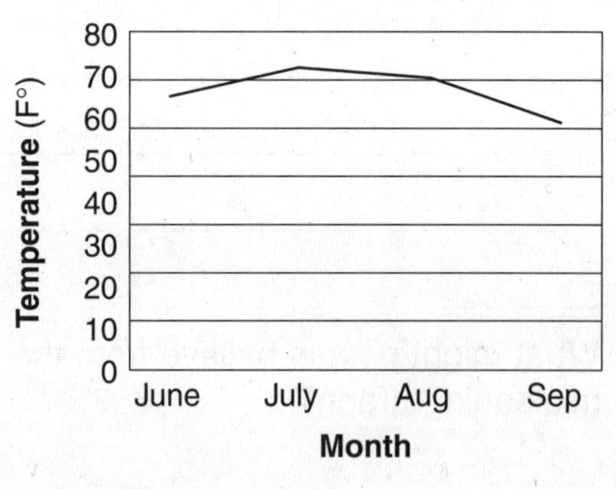

3. Why are these line graphs misleading?

4. What might people believe from the misleading graphs?

Name ________________________ Date __________ Class __________

LESSON 6-8

Reteach

Misleading Graphs

Graphs are often made to influence you. When you look at a graph, you need to figure out if the graph is accurate or if it is misleading.

Look at the graph below.

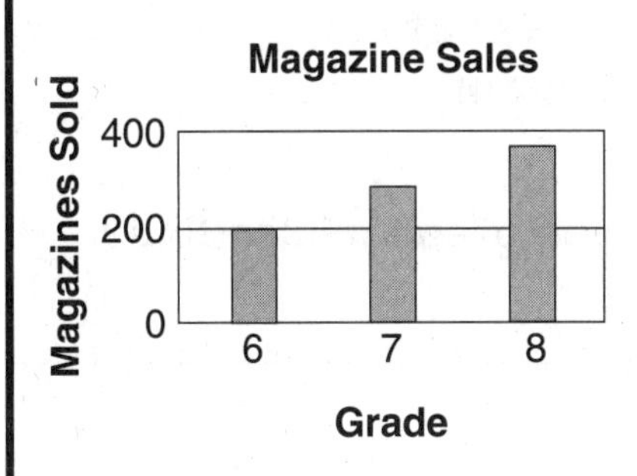

The graph is misleading because the intervals for the scale are so great. When you first look at the graph it appears that each grade sold about the same number of magazines.

Look at each graph. Then explain why each graph is misleading.

1.

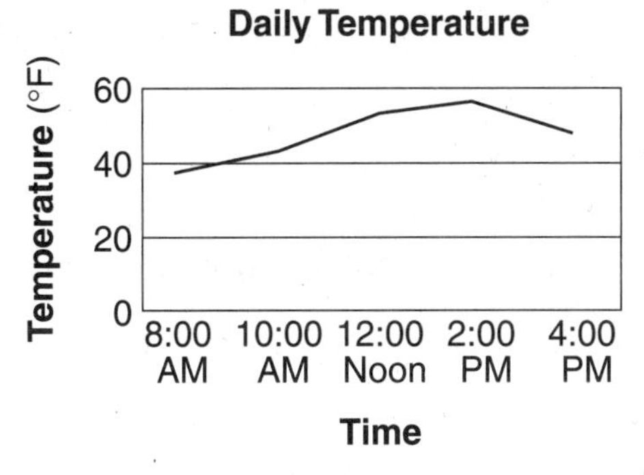

____
____
____
____
____
____

2.

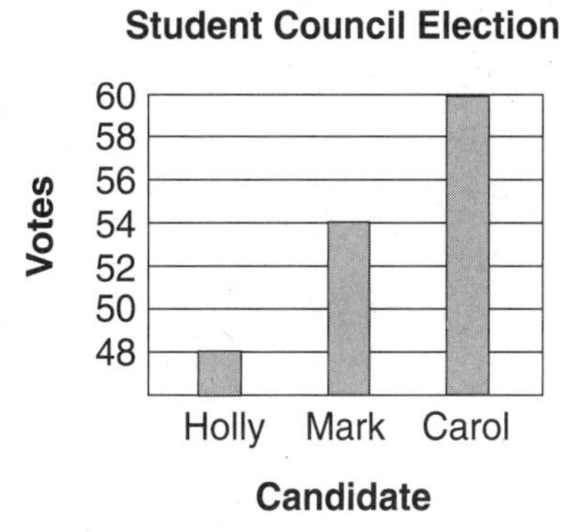

____
____
____
____
____
____

Name ______________________ Date ____________ Class ____________

LESSON 6-8

Challenge

Graph Detective

You are a police detective in Capital City. A gang of criminals there is distributing misleading graphs to convince people that your city does not need to hire more police officers. It's your job to catch these graph crooks.

Search the graphs below for evidence of misleading displays of data. Then use your detective skills to explain why each graph is misleading.

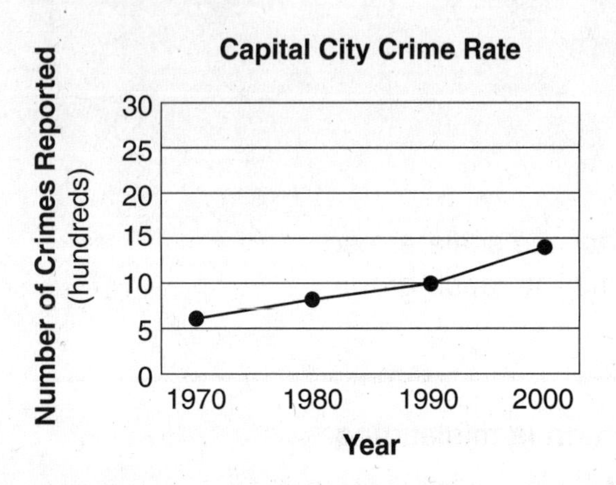

Why is this line graph misleading?

____

____

____

What might people believe from this misleading graph?

____

____

____

____

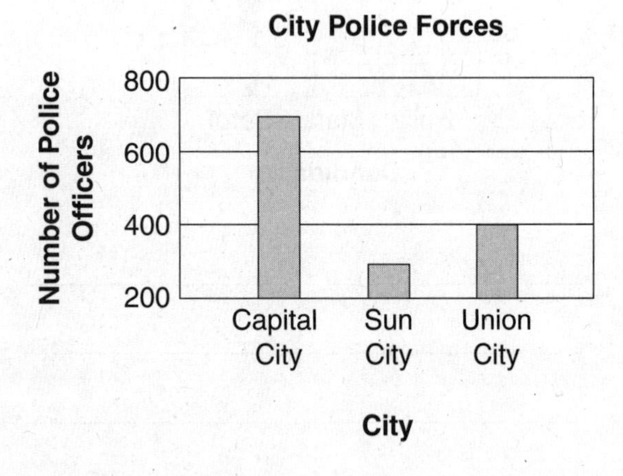

Why is this bar graph is misleading?

____

____

____

____

What might people believe from this misleading graph?

____

____

____

____

____

Name ______________________ Date __________ Class __________

LESSON 6-8

Problem Solving

Misleading Graphs

Use the graphs to answer each question.

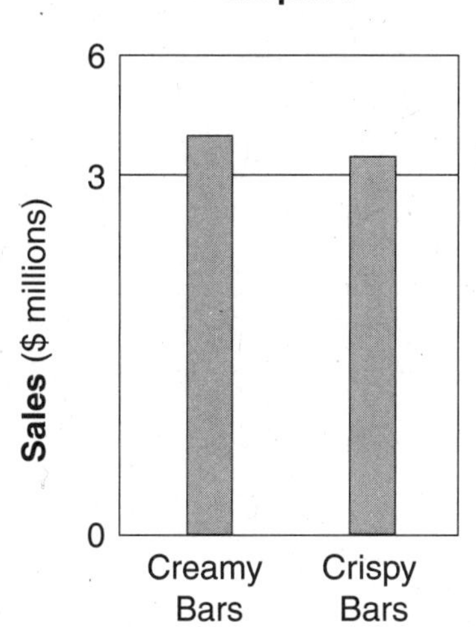

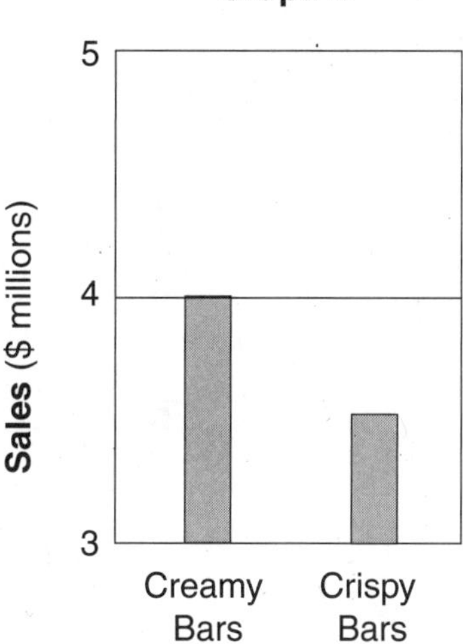

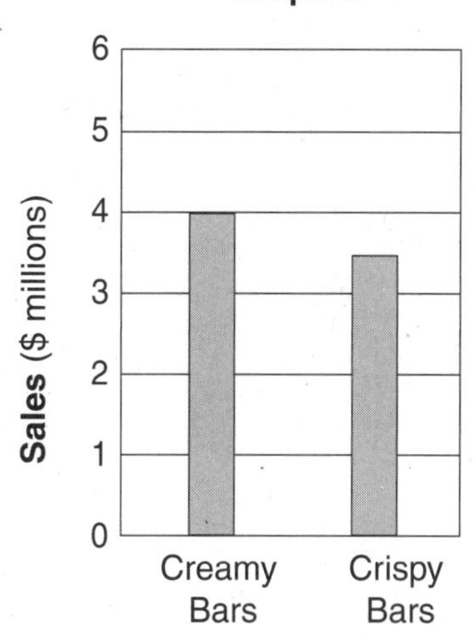

1. Why is Graph A misleading?

____

____

____

2. Why is Graph B misleading?

____

____

____

3. What might people believe from reading Graph A?

____

____

4. What might people believe from reading Graph B?

____

____

Circle the letter of the correct answer.

5. Which of the following information is different on all three graphs above?
   A the vertical scale
   B the Crispy Bars sales data
   C the Creamy Bars sales data
   D the horizontal scale

6. Which of the following is a way that graphs can be misleading?
   F breaks in scales
   G uneven scales
   H missing parts of scales
   J all of the above

7. Which graph do you think was made by the company that sells Crispy Bars?
   A Graph A
   B Graph B
   C Graph C
   D all of the graphs

8. If you were writing a newspaper article about candy bar sales, which graph would be best to use?
   F Graph A
   G Graph B
   H Graph C
   J all of the above

Name ______________________ Date ____________ Class ____________

LESSON 6-8

Reading Strategies

Compare and Contrast

A graph can be misleading if the scale of the graph does not start at 0. Compare Graph A and Graph B.

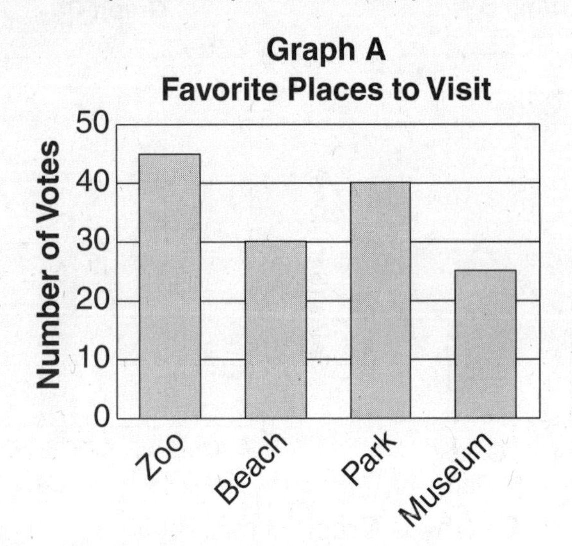

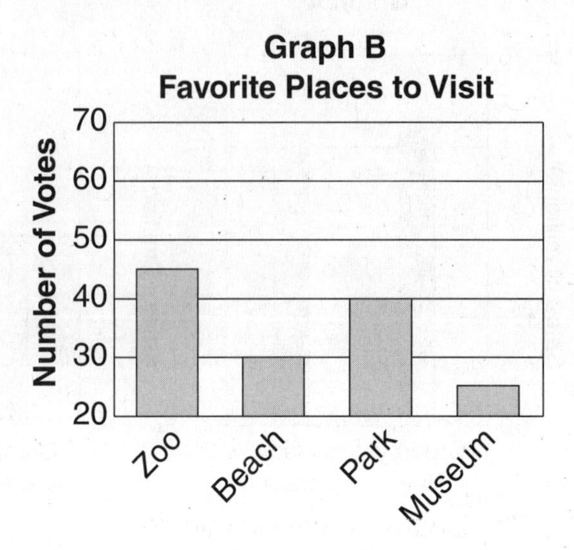

Answer these questions about the graphs.

1. What are the titles of Graph A and Graph B?

____

2. What do the numbers on the left side of the graph show?

____

3. By how much do the number of votes increase along the left side of the graph?

____

4. What do the bars on the graph stand for?

____

5. What is the first number shown along the left side of Graph A? ____________

6. What is the first number shown along the left side of Graph B? ____________

7. Compare the bars on Graph A to the bars on Graph B.

____

8. Which graph is misleading?

____

Name ______________________ Date __________ Class __________

LESSON **6-8**

Puzzles, Twisters & Teasers

Doctor, Doctor

Two doctors are advertising in the paper. Here are their newspaper advertisements.

Graph #1

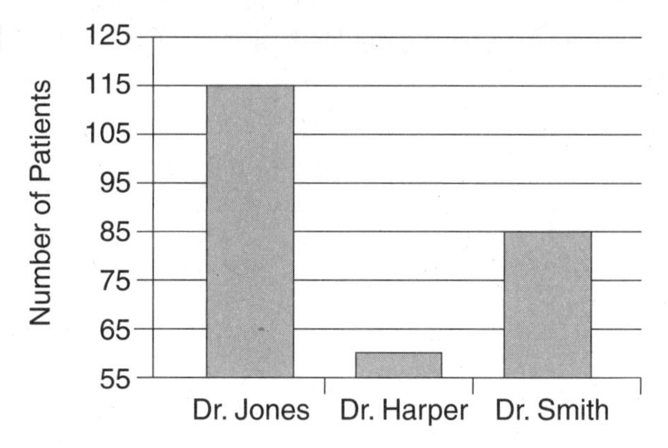

"Come to Dr. Jones. I have so many more patients than the other doctors in town because people trust me."

Graph #2

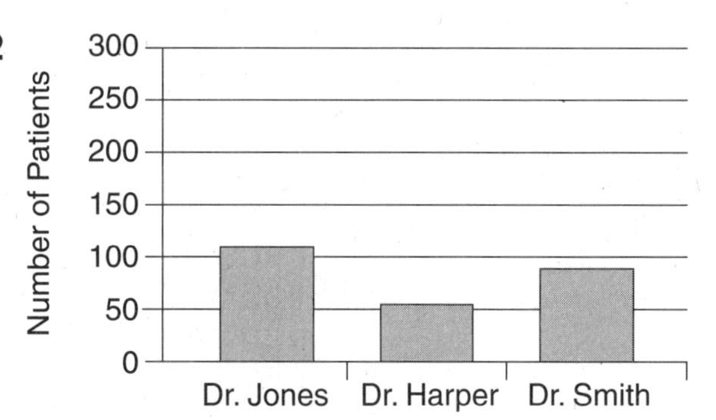

"Come to Dr. Smith. I just graduated and know all the most recent medical procedures."

Answer the questions to solve the riddle:

1. In which graph does it look like Doctor Jones has about the same number of patients as other doctors? _____ **P**
2. By what number do the numbers increase on the *y*-axis in graph #2? _____ **N**
3. By what number do the numbers increase on the *y*-axis in graph #1? _____ **A**
4. What is the highest number on the *y*-axis on graph #2? _____ **I**
5. About how many patients does Doctor Jones have? _____ **T**
6. About how many more patients does Doctor Jones see than Dr. Smith? _____ **E**
7. If you were Dr. Jones, which graph would you want people to see? _____ **T**

A man runs into the doctor's office and says, "Hey Doc, I think I'm shrinking. Do something right away!"

The doctor replies:
"YOU'LL JUST HAVE TO BE A LITTLE ___ ___ ___ ___ ___ ___ ___."

2 10 115 300 30 50 1

Name ______________________ Date ____________ Class ____________

LESSON 6-9

Practice A

Stem-and-Leaf Plots

Complete each activity and answer each question.

1. Use the data in the table to complete the stem-and-leaf plot below.

Daily Low Temperatures (°F)	16	21	15	27	30	25

Daily Low Temperature

Stem	Leaves

Key: 1 | 6 = ____________

Find each value of the data.

2. smallest value ____________

3. largest value ____________

4. mean ____________

5. median ____________

6. mode ____________

7. range ____________

Stem	Leaves
1	0 5
2	3
3	7
4	5

Key: 2 | 4 = 24

8. In the stem-and-leaf plot for Exercises 2–7, which digit was used for the stems? for the leaves?

____

9. Look at the stem-and-leaf plot you made for Exercise 1. What are the smallest and largest values in the data set?

____

Name ______________________ Date __________ Class __________

LESSON 6-9

Practice B

Stem-and-Leaf Plots

Complete each activity and answer the questions.

1. Use the data in the table to complete the stem-and-leaf plot below.

Richmond, Virginia, Monthly Normal Temperatures (°F)											
Jan	Feb	Mar	April	May	June	July	Aug	Sep	Oct	Nov	Dec
37	39	48	57	74	78	77	76	70	59	50	40

Stem	Leaves

Key: 1 | 2 = __________

Find each value of the data.

2. least value __________

3. greatest value __________

4. mean __________

5. median __________

6. mode __________

7. range __________

Stem	Leaves
6	1 4
7	1 6
8	2 2
9	0 1 8

Key: 6 | 5 = 65

8. Look at the stem-and-leaf plot you made for Exercise 1. How many months in Richmond have a normal temperature above 70°F?

____

____

9. How would you display a data value of 100 on the stem-and-leaf plot above?

____

____

Name ______________________ Date ___________ Class ___________

LESSON 6-9

Practice C

Stem-and-Leaf Plots

Complete each activity and answer the questions.

1. Use the data in the table to make a stem-and-leaf plot.

Books Read by Read-A-Thon Participants									
50	19	24	45	44	12	32	19	38	43
35	40	15	19	26	30	28	40	12	18

Stem	Leaves

Key: 1 | 2 = ___________

Find each value of the data.

2. least value ___________

3. greatest value ___________

4. mean ___________

5. median ___________

6. mode ___________

7. range ___________

Stem	Leaves
1	2 5 8 9
2	0 4 5 6 8 8
3	2 6
5	0 2

Key: 3 | 3 = 33

8. Look at the stem-and-leaf plot you made for Exercise 1. How many students read more than 40 books during the read-a-thon?

____

____

9. How would you display a data value of 5 on the stem-and-leaf plot above? What would be the mean of this new data set?

____

____

Name ______________________ Date __________ Class __________

CHAPTER 6-9

Reteach

Stem-and-Leaf Plots

You can use place value to make a stem-and-leaf plot.

Points Earned in Games During Basketball Season						
27	16	34	29	48	12	33
20	18	42	51	27	32	41

Write the numbers in order from least to greatest.

12 16 18 20 27 27 29 32 33 34 41 42 48 51

List the tens digits in order from least to greatest in the first, or stem, column. Then, for each tens digit, record the ones digit for each data value in order from least to greatest in the second, or leaves, column.

Points Earned

Stem	Leaves
1	2 6 8
2	0 7 7 9
3	2 3 4
4	1 2 8
5	1

Key: 1 | 2 = 12

Make sure your graph has a title and a key.

Use the data to make a stem-and-leaf plot.

1.

Valerie's Test Scores				
62	84	93	88	89
76	68	81	91	88

Valerie's Test Scores

Stem	Leaves

Key: 6 | 2 = __________

2. What is the range?

____

____

3. What is the median?

____

____

4. What is the mode?

____

____

Name ______________________ Date ___________ Class ___________

LESSON 6-9

Challenge

A Plot of Trees

As part of your job with the National Forest System, you must compile a report on the tallest trees in the United States. You have already collected the data and displayed it on the bar graph below. To complete the report, you need to make a stem-and-leaf plot of the data. Use the hundreds digits of the data for your stems. Then analyze the data.

National Forest System Report, 2000

Tallest Trees in the U.S.	**Data Analysis**
Stem | **Leaves** Key: 1 | 78 = __________	Range of heights: __________ Mean height: __________ Median height: __________ Mode height: __________

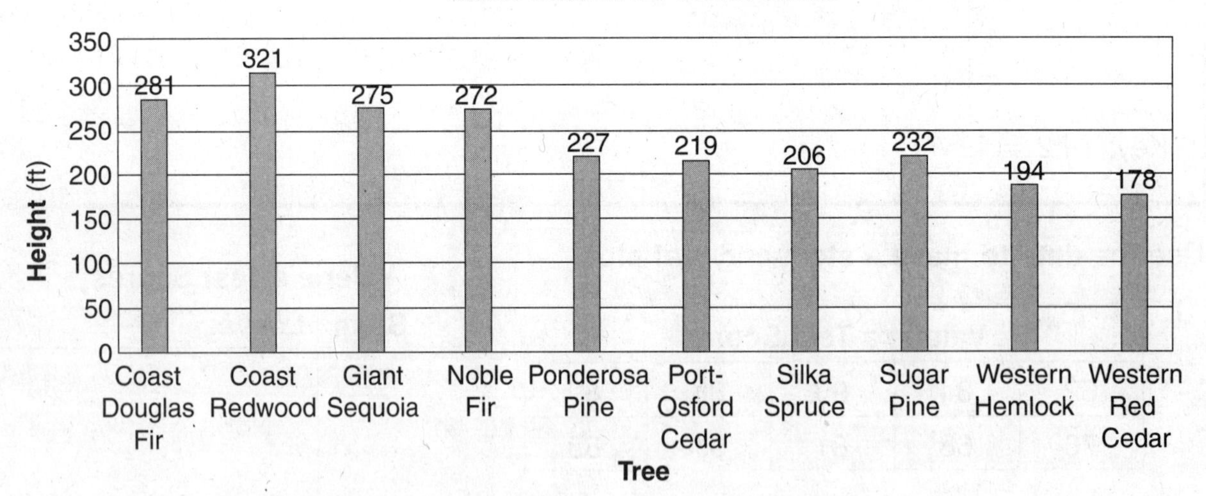

Name ______________________________ Date ____________ Class ____________

LESSON 6-9

Problem Solving

Stem-and-Leaf Plots

Use the Texas stem-and-leaf plots to answer each question.

Dallas Normal Monthly Temperatures

Stem	Leaves
4	3 7 8
5	6 7
6	6 7
7	3 7
8	1 5 5

Key: 4 | 3 = 43°F

Houston Normal Monthly Temperatures

Stem	Leaves
5	0 4 4
6	1 1 8
7	0 5 8
8	0 2 3

Key: 5 | 0 = 50°F

1. Which city's temperature data has a mode of 85°F?

____

2. Which city's temperature data has a range of 33°F?

____

3. Which city has the lowest data value? What is that value?

____

4. Which city has the highest data value? What is that value?

____

Circle the letter of the correct answer.

5. Which city's temperature data has a mean of 68°F?

A Dallas

B Houston

C both Dallas and Houston

D neither Dallas nor Houston

6. Which city's temperature data has a median of 69°F?

F Dallas

G Houston

H both Dallas and Houston

J neither Dallas nor Houston

7. What do the data values 54°F and 61°F represent for the plots above?

A the ranges of normal temperatures in Dallas and Houston

B the mode of normal temperatures for Houston

C the mean and median normal temperatures for Dallas

D the lowest normal temperatures for Dallas and Houston

8. Which of the following would be the best way to display the Dallas and Houston temperature data?

F on a line graph

G in a tally table

H on a bar graph

J on a coordinate plane

Name ______________________ Date ____________ Class ____________

LESSON 6-9

Reading Strategies

Use a Graphic Organizer

Below is a list of high temperatures during a two-week period in Austin, Texas.

75 78 63 79 74 73 83 72 85 62 84 65 68 81

Making a table is one way to organize the temperature data so it is easier to understand.

63	75	83
62	78	85
65	79	84
68	74	81
	72	
	73	

Answer each question about the table.

1. How were the temperatures organized in the table?

____

____

2. How many days was the temperature in the 70's?

____

3. How many days was the temperature above 80? ____________________

4. Complete: Temperatures were mostly in the ____________ during this two-week period.

5. Put the temperatures with 8 in the tens place in order from least to greatest.

____

6. What was the range of high temperatures during the two-week period?

____

7. How did organizing the temperatures help you answer the questions above?

____

____

Name ______________________ Date __________ Class __________

Puzzles, Twisters & Teasers

LESSON 6-9

It's a Leaf

Arrange the test scores below in a stem-and-leaf plot from least to greatest. After organizing the data in the plot, count and record the number of leaves per stem. Then find the letter matching the number of leaves. Solve the riddle by arranging the number of leaves from least to greatest.

Test Scores: 56, 70, 100, 65, 48, 92, 84, 95, 97, 100, 68, 75, 81, 85
59, 92, 96, 66, 75, 83, 66, 72, 85, 93, 73, 95, 100, 87

Stem	Leaves

Number of Leaves	Letter
1	

Letters

A = 8	**D** = 10	**I** = 3	**L** = 6	**S** = 13
B = 11	**E** = 0	**J** = 9	**N** = 12	**U** = 2
C = 4	**G** = 14	**K** = 5	**Q** = 1	**Y** = 7

How do hikers dress on cold mornings?

___ ___ ___ ___ ___ ___ ___

Name ______________________ Date ____________ Class ____________

LESSON 6-10

Practice A

Choosing an Appropriate Display

1. The table shows the average high temperatures in San Diego for six months of one year. Which graph would be more appropriate to show the data—a line graph or a bar graph? Draw the more appropriate graph.

Month	Mar	June	Aug	Sep	Oct	Dec
Average High Temperature (°F)	65	70	77	76	73	64

2. The table shows the results of a survey about students' favorite snack. Which graph would be more appropriate to show the data—a line graph or a bar graph? Draw the more appropriate graph.

Type of Snack	Popcorn	Fruit	Cheese	Pretzels
Number of Votes	16	6	2	9

Name ______________________ Date __________ Class __________

LESSON 6-10

Practice B

Choosing an Appropriate Display

1. The table shows the heights of the 6 tallest buildings in the world. Which graph would be more appropriate to show the data—a line graph or a bar graph? Draw the more appropriate graph.

Building	Sears Tower	CITIC Plaza	Petronas Tower I	Petronas Tower II	Jin Mao Building	Two Finance Center
Height (ft)	1,450	1,283	1,483	1,483	1,381	1,352

2. The table shows the test scores of some sixth-grade students. Which graph would be more appropriate to show the data—a stem-and-leaf plot or a line graph? Draw the more appropriate graph.

Test Scores												
62	78	81	66	96	88	81	77	90	88	60	99	90

Name ______________________ Date ____________ Class ____________

LESSON 6-10

Practice C

Choosing an Appropriate Display

1. The table shows the populations of the 6 largest Native American tribes in the United State. Which graph would be more appropriate to show the data—a line graph or a bar graph? Draw the more appropriate graph.

Tribe	**Navajo**	**Sioux**	**Chippewa**	**Cherokee**	**Choctaw**	**Latin American Indian**
Population	269,202	108,272	105,907	281,069	87,349	104,354

2. The table shows the heights of some sixth-grade students. Make a graph of the data, using the most appropriate way to display it.

Heights of Sixth-Graders (in.)												
52	55	60	53	51	54	55	51	53	50	50	52	51

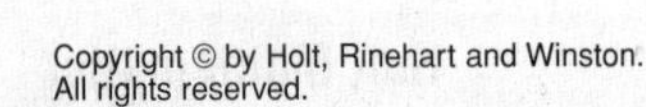

Name ______________________ Date __________ Class __________

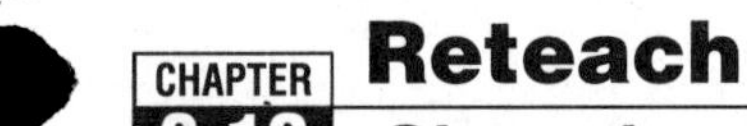

Reteach

Choosing an Appropriate Display

You can choose the best way to display data by thinking about the data you want to show.

Use a line plot to show the frequency of data on a number line.

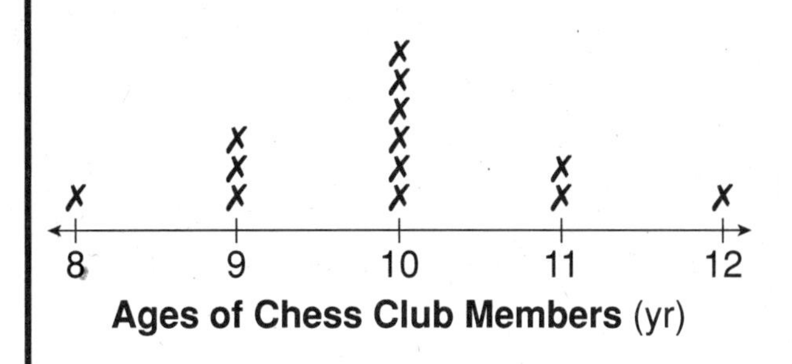

Use a line graph to show changes in data over time.

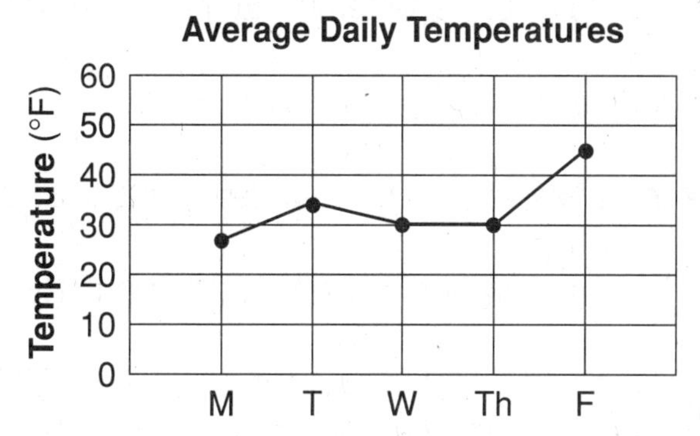

Use a bar graph to display data in separate categories.

Use a stem-and-leaf plot to show how often data values occur and how they are distributed.

Test Scores

Stem	Leaves
7	0 0 1 3 8
8	1 2 5 8
9	0 0 4 5 5 7

Key: 7|0 = 70

Display the data. Use the most appropriate type of graph.

Month	Mar	Apr	May	June	July
Average Temperature (°C)	12	18	20	23	25

Name ______________________ Date __________ Class __________

LESSON 6-10

Challenge

Graphing the News

Your job as a television news reporter is to make data displays for use on each evening's newscast. Tell the most appropriate type of graph to use for each kind of data described below.

- the daily average temperature for five days ____________________
- the six most popular movies last month ____________________
- the ages of twenty world leaders ____________________
- the amount of money collected by thirty charity organizations ____________________

Choose one of the kinds of data above and write a data set for it. Then prepare the graph for this evening's news telecast.

Name ______________________ Date __________ Class __________

LESSON 6-10

Problem Solving

Choosing an Appropriate Display

1. Write *line plot, stem-and-leaf plot, line graph,* or *bar graph* to describe the most appropriate way to show the height of a sunflower plant every week for one month.

   ____

2. Write *line plot, stem-and-leaf plot, line graph,* or *bar graph* to describe the most appropriate way to show the number of votes received by each candidate running for class president

   ____

3. Write *line plot, stem-and-leaf plot, line graph,* or *bar graph* to describe the most appropriate way to show the test scores each student received on a math quiz.

   ____

4. Write *line plot, stem-and-leaf plot, line graph,* or *bar graph* to describe the most appropriate way to show the average time spent sleeping per day by 30 sixth-grade students.

   ____

Circle the letter of the correct answer.

5. People leaving a restaurant were asked how much they spent for lunch. Here are the results of the survey to the nearest dollar: \$8, \$7, \$9, \$7, \$10, \$5, \$8, \$8, \$12, \$8. Which type of graph would be most appropriate to show the data?

   A bar graph
   B line graph
   C line plot
   D stem-and-leaf plot

6. People leaving a movie theater were asked their age. Here are the results of the survey to the nearest year: 12, 11, 13, 15, 22, 31, 40, 12, 17, 20, 33, 16, 12, 24, 19. Which type of graph would be most appropriate to show the data?

   F bar graph
   G line graph
   H line plot
   J stem-and-leaf plot

7. What is the median amount of money spent on lunch in Exercise 5?

   A \$7
   B \$8
   C \$9
   D \$12

8. What is the median age of the movie-goers in Exercise 6?

   F 15
   G 16
   H 17
   J 19

Name ______________________ Date ____________ Class ____________

LESSON 6-10

Reading Strategies

Understand Vocabulary

A **line plot** is a number line with marks or dots that show frequency of data.

A **bar graph** is a graph that uses vertical or horizontal bars to display data in discrete categories.

A **line graph** is a graph that uses line segments to show how data changes over time.

A **stem-and-leaf plot** is a graph used to organize and display data so that the frequencies can be compared.

Use the definitions to help you answer each question.

1. Which type of display would be best to use to show the data in the table at right?

____

Type of Club	Number of Members
chess	12
crafts	23
music	14
reading	19

2. Which type of display would be best to use to show the data in the table at right?

____

Month	Average Rainfall (in.)
February	1.2
March	2.4
April	3.8
May	2.1

3. Which type of display would be best to use to show the data in the table at right?

____

Ages of Club Members					
11	12	11	10	11	12
13	10	14	11	10	9
10	11	12	12	11	11
11	11	10	11	10	10

Name ______________________ Date __________ Class __________

LESSON 6-10

Puzzles, Twisters & Teasers

Way to Display!

Yoko, Ken, Olivia, and Larry want to make data displays. Use the following descriptions to match the student to the best type of display for his or her data.

- Yoko wants to display the number of kilometers several people ran in last week's run for charity.
- Ken wants to display the History Museum attendance over a period of six months.
- Olivia wants to display the science test scores of all of her classmates.
- Larry wants to display the results of a survey he took about some friends' favorite weekend activities.

	1 LINE PLOT	2 STEM-AND-LEAF PLOT	3 BAR GRAPH	4 LINE GRAPH
Ken				
Olivia				
Yoko				
Larry				

To solve the riddle, fill the spaces above each number with the first letter of the name of the person who will make that display number.

What do you call a funny book about eggs?

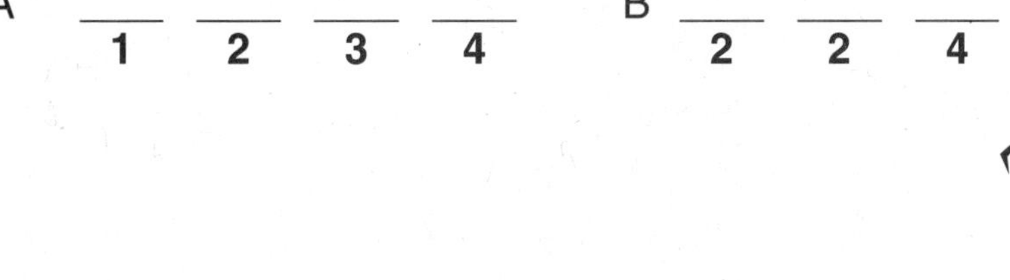

A ___ ___ ___ ___
1 2 3 4

B ___ ___ ___
2 2 4

LESSON 6-1

Practice A

Problem Solving Skill: Make a Table

Complete each activity and answer the questions.

1. There are five children in the Cooper family. July is 8 years old, Brent is 16 years old, Michael is 10 years old, Andrea is 14 years old, and Matthew is 12 years old. Use this data to complete the table at right showing the children's ages in order from youngest to oldest.

Cooper Children Ages

Child	Age
July	8
Michael	10
Matthew	12
Andrea	14
Brent	16

2. What pattern do you see in the table's data?

The ages increase by two years between each child.

3. On Monday, the low temperature was 41°F. On Tuesday, the low temperature was 38°F. On Wednesday, the low temperature was 35°F. On Thursday, the low temperature was 32°F and on Friday the low temperature was 29°F. Use this data to complete the table at right.

Daily Low Temperatures

Day	Temperature (°F)
Monday	41
Tuesday	38
Wednesday	35
Thursday	32
Friday	29

4. What pattern do you see in the table's data?

Each day, the temperature dropped by 3°F.

5. Look at the table you completed for Exercise 1. Name a different way that you could organize the same data in the table.

Possible answers: by age, from oldest to youngest; by name, in alphabetical order

6. Look at the table you completed for Exercise 3. Name a different way that you could organize the same data in the table.

Possible answers: by temperature, from hottest to coldest or from coldest to hottest

LESSON 6-1

Practice B

Problem Solving Skill: Make a Table

Complete each activity and answer each question.

1. Pizza Express sells different-sized pizzas. The jumbo pizza has 20 slices. The extra large has 16 slices. The large has 12 slices. There are 8 slices in a medium, and 6 slices in a small. A personal-sized pizza has 4 slices. Use this data to complete the table at right, from largest to smallest pizza.

Pizza Express Pizza Sizes

Size	Slices
Jumbo	20
Extra large	16
Large	12
Medium	8
Small	6
Personal	4

2. What pattern do you see in the table's data?

The first 4 sizes decrease by 4 slices each; the last 3 sizes decrease by 2 slices each.

3. A plain large pizza at Pizza Express costs $13.75. A large pizza with one topping costs $14.20. A 2-topping large pizza costs $14.65, and a 3-topping large pizza costs $15.10. If you want 4 toppings on your large pizza, it will cost you $15.55. Use this data to complete the table at right.

Pizza Express Pizza Prices

Large Pizza	Price
Plain	$13.75
1 Topping	$14.20
2 Toppings	$14.65
3 Toppings	$15.10
4 Toppings	$15.55

4. What pattern do you see in the table's data?

The price increases by $0.45 for each additional topping.

5. How much does each slice of a 1-topping large pizza from Pizza Express cost? Round your answer to the nearest hundredth of a dollar.

$1.18 per slice

6. You and three friends buy two large pizzas from Pizza Express. One pizza has pepperoni and onions, and one pizza is plain. If you equally share the total price, how much will you each pay? How many slices will you each get?

$7.10 each; 6 slices each

LESSON 6-1

Practice C

Problem Solving Skill: Make a Table

Complete each activity and answer each question.

1. At 9:00 A.M., the temperature in Yuma, Arizona, was 79°F. At noon, the temperature was 83°F. At 2:00 P.M., the temperature was 87°F. At 4:00 P.M., the temperature was 84°F. At 6:00 P.M., the temperature was 81°F. By 8:00 P.M., the temperature in Yuma was 78°F. Use this data to complete the two tables below. Each table's data should be organized differently.

Possible organization of data in tables is given.

Table A

Time	Temperature (°F)
9:00 A.M.	79
12:00 P.M.	83
2:00 P.M.	87
4:00 P.M.	84
6:00 P.M.	81
8:00 P.M.	78

Table B

Time	Temperature (°F)
8:00 P.M.	78
9:00 A.M.	79
6:00 P.M.	81
12:00 P.M.	83
4:00 P.M.	84
2:00 P.M.	87

2. Use Table A to find a pattern in the data and draw a conclusion.

Pattern: The temperature increased by 4°F with each of the first 3 readings, and decreased by 3°F with each of the last 3. Possible conclusion: It was 75°F at both 7:00 A.M. and 10:00 P.M.

3. Use Table B to find a pattern in the data and draw a conclusion.

Pattern: The 2 lowest temperatures were at the beginning and end of the day. Possible conclusion: The lowest temperature on this day in Yuma was at least 78°F.

4. When would each of these tables be most useful for finding and comparing data?

Possible answers: Table A for finding how the temperature changed throughout the day; Table B for finding the highest and lowest temperatures

LESSON 6-1

Reteach

Problem Solving Skill: Make a Table

You can make a table to organize data. Then you can use the table to see patterns and draw conclusions.

During the week-long book fair, 324 books were sold. On Monday, 45 books were sold. On Tuesday, students bought 58 books. On Wednesday, 79 books were sold. Sixty-two books were sold on Thursday, and students bought 51 books on Friday.

Day	Books Sold
Monday	45
Tuesday	58
Wednesday	79
Thursday	62
Friday	51

To make a table, arrange the information in order by days so you can see patterns over time. Remember to make headings for each column of the table.

From the table, you can see that the number of books sold increased from Monday to Wednesday, and decreased from Wednesday to Friday.

Use the data to make a table. Then use the table to find a pattern in the data and draw a conclusion.

1. During the championship series, the school basketball team earned 24 points in the first game, 28 points in the second game, 33 points in the third game, 42 points in the fourth game, and 49 points in the last game.

Game	Points Earned
1	24
2	28
3	33
4	42
5	49

The team earned more points each time that it played a game. The team earned at least 24 points each game.

2. In the sixth grade, 18 students study Spanish, 35 students study French, 11 students study Latin, and 5 students study no foreign language at all.

Foreign Language	Students
French	35
Spanish	18
Latin	11
None	5

Most of the sixth grade students study French as a foreign language.

LESSON 6-1

Challenge
Liberty Logic

You can use tables and logic to organize information and solve problems. For example, you have some measurements for different parts of the Statue of Liberty, but you do not know which measurement goes with which part. To solve the problem, first organize all the possibilities in a logic table. Then use the clues to fill out the table.

Because each part has only one measurement, there can be only one **YES** in each row and column of your logic table.

Three measurements for parts of the statue's face are 30 inches, 36 inches, and 54 inches. Which of those measurements are for the width of her mouth, the length of her nose, and the width of each of her eyes?

Clue 1: Her mouth is wider than each of her eyes.
Clue 2: The length of her nose is greater than the width of her mouth.

	30 inches	36 inches	54 inches
Width of Mouth	NO	YES	NO
Length of Nose	NO	NO	YES
Width of each Eye	YES	NO	NO

Three measurements for the tablet she holds are 24 inches, 163 inches, and 283 inches. Use the clues and logic table below to find the length, width, and thickness of the Statue of Liberty tablet.

Clue 1: The tablet is longer than it is wide.
Clue 2: The tablet is less than 100 inches thick.

	24 inches	163 inches	283 inches
Length	NO	NO	YES
Width	NO	YES	NO
Thickness	YES	NO	NO

LESSON 6-1

Problem Solving
Problem Solving Skill: Make a Table

Complete each activity and answer each question.

1. In January, the normal temperature in Atlanta, Georgia, is 41°F. In February, the normal temperature in Atlanta is 45°F. In March, the normal temperature in Atlanta is 54°F, and in April, it is 62°F. Atlanta's normal temperature in May is 69°F. Use this data to complete the table at right.

Atlanta Normal Temperatures

Month	Temperature (°F)
January	41
February	45
March	54
April	62
May	69

2. Use your table from Exercise 1 to find a pattern in the data and draw a conclusion about the temperature in June.

Pattern: The normal temperature in Atlanta increases each month from January to May. Possible conclusion: Atlanta's normal temperature in June is higher than 69°F.

3. In what other ways could you organize the data in a table?

Possible answers: by temperature from lowest to highest, or from highest to lowest

Circle the letter of the correct answer.

4. In which month given does Atlanta have the highest temperature?
   - A February
   - B March
   - C April
   - (D) May

5. In which month given does Atlanta have the lowest temperature?
   - (F) January
   - G February
   - H March
   - J April

6. Which of these statements about Atlanta's temperature data from January to May is true?
   - (A) It is always higher than 40°F.
   - B It is always lower than 60°F.
   - C It is hotter in March than in April.
   - D It is cooler in February than in January.

7. Between which two months in Atlanta does the normal temperature change the most?
   - F January and February
   - (G) February and March
   - H March and April
   - J April and May

LESSON 6-1

Reading Strategies
Reading a Table

Jill jumps rope each day as part of her fitness program. She made a **table** to keep track of how much time she spent jumping rope each day for one week.

Day	Time
Monday	15 minutes
Tuesday	18 minutes
Wednesday	18 minutes
Thursday	21 minutes
Friday	21 minutes
Saturday	24 minutes
Sunday	24 minutes

1. What headings are shown in the table? day and time
2. What does "time" stand for in the table? time spent jumping rope
3. How long did Jill jump rope on Tuesday? 18 minutes
4. What other day of the week did Jill jump rope for the same length of time? Wednesday
5. On which day did Jill jump rope for 15 minutes? Monday
6. List in order from least to greatest the different times Jill spent jumping rope. 15 min, 18 min, 21 min, 24 min
7. What pattern do you notice in this list of times? Each time increases by 3 minutes.
8. If the pattern continues, how much time would you expect Jill to jump rope on the following Monday? 27 minutes
9. How did the table help you understand Jill's fitness program? Possible answer: Having the data organized made it easy to answer the questions and see patterns in the data.

LESSON 6-1

Puzzles, Twisters & Teasers
Making the Grade

Charlie, Pat, Oliver, Eduardo, and Hannah each received different grades from B all the way up to an A+ on the last test. Given the following clues, use the yes/no table to determine the grade each of the students received.

1. Oliver received some sort of A.
2. Hannah and Pat each got a higher grade than Eduardo.
3. Charlie got the A−.
4. Pat's grade was higher than Oliver's was.

	A+	A	A−	B+	B
Hannah					NO(#2)
Eduardo	NO(#2)				
Oliver	NO(#4)		NO	NO	NO
Pat					NO(#2)
Charlie			YES(#3)		

To solve the riddle, fill the spaces above each of the grades with the first letter of the name of the person who received that grade.

Implies:

	A+(P)	A (O)	A- (C)	B+ (H)	B (E)
Hannah	NO	NO	NO	YES	NO(#2)
Eduardo	NO(#2)	NO	NO	NO	YES
Oliver	NO(#4)	YES	NO(#1)	NO(#1)	NO(#1)
Pat	YES	NO	NO	NO	NO(#2)
Charlie	NO	NO	YES(#3)	NO	NO

What do you get when you cross a dog and a hen?

P (A+) O (A) O (A) C (A−) H (B+) E (B) D E (B) G G S

LESSON 6-2 **Practice A**

Mean, Median, Mode, and Range

Find the mean of each data set.

1.

Length of Worms (in.)	3	5	4	2	6

mean: 4 in.

2.

Ages of Brothers (yr)	12	16	15	14	8

mean: 13 y

Find the mean, median, mode, and range of each data set.

3.

Heights of Trees (m)	7	11	9	7	6

mean: 8 m; median: 7 m; mode: 7 m; range: 5 m

4.

Sizes of Bottled Juice (L)	6	12	12	16	24

mean: 14 L; median: 12 L; mode: 12 L; range: 18 L

5.

Football Team Wins (games per season)	10	8	10	8	14

mean: 10 games; median: 10 games; mode: 8 and 10 games; range: 6 games

6. Tammy is 14 years old. She has a younger sister and an older brother. Her sister is 12 years old. The mean of all their ages is 14. How old is Tammy's brother?

16 years old

7. The mode of Nevin's four math quiz scores last month is 85 points. On three of the quizzes, he earned the following scores: 90, 86, and 85. What was the score of Nevin's other quiz?

85 points

LESSON 6-2 **Practice B**

Mean, Median, Mode, and Range

Find the mean of each data set.

1.

Brian's Math Test Scores	86	90	93	85	79	92

mean: 87.5

2.

Heights of Basketball Players (in.)	72	75	78	72	73

mean: 74 in.

Find the mean, median, mode, and range of each data set.

3.

School Sit-Up Records (sit-ups per minute)	31	28	30	31	30

mean: 30 sit-ups; median: 30 sit-ups; mode: 30 and 31 sit-ups; range: 3 sit-ups

4.

Team Heart Rates (beats per min)	70	68	70	72	68	66

mean: 69 bpm; median: 69 bpm; mode: 68 and 70 bpm; range: 6 bpm

5.

Daily Winter Temperatures (°F)	45	50	47	52	53	45	51

mean: 49°F; median: 50°F; mode: 45°F; range: 8°F

6. Anita has two sisters and three brothers. The mean of all their ages is 6 years. What will their mean age be 10 years from now? Twenty years from now?

16 years; 26 years

7. In a class of 28 sixth graders, all but one of the students are 12 years old. Which two data measurements are the same for the student's ages? What are those measurements?

the median and mode; 12 years

LESSON 6-2 **Practice C**

Mean, Median, Mode, and Range

Find the mean of the data set.

1.

Monthly Girl Scout Cookie Sales (boxes per person)									
22	13	47	11	8	16	15	14	13	17

mean: 17.6 boxes

Find the mean, median, mode, and range of each data set.

2.

Monthly Rainfall (in.)	7.6	6.7	8.1	6.2	6.0	6.2

mean: 6.8 in.; median: 6.45 in.; mode: 6.2 in.; range: 2.1 in.

3.

Wildcat Basketball Season Wins (number of points won by)							
24	12	10	18	20	12	17	10

mean: 15.375 pt; median: 14.5 pt; mode: 10 and 12 pt; range: 14 pt

4.

Tom's Weekly Earnings ($)	200	167	185	212	195	193

mean: $192; median: $194; mode: none; range: $45

5. There are seven whole numbers in a data set. The mean of the data set is 28. The median is 29, and the mode is 31. The least number in the data set is 22, and the greatest number is 35. What are the seven numbers in the data set?

Possible answer: 22, 23, 25, 29, 31, 31, and 35

6. There are seven children in the Arthur family, including one set of twins. The youngest child is 6 years old, and the oldest is 16 years old. The mean of their ages is 11 years, the median is 10 years, and the mode is 15 years. What are the ages of the Arthur children? How old are the twins?

6, 7, 8, 10, 15, 15, and 16 years; 15 years old

LESSON 6-2 **Reteach**

Mean, Median, Mode, and Range

You can find the mean, median, mode, and range to describe a set of data.

Terry's Test Scores	76	81	94	81	78

The **mean** or average is the sum of the items divided by the number of items.

$76 + 81 + 94 + 81 + 78 = 410$ First, find the sum of the values.

$410 \div 5 = 82$ Then divide the sum by the number of values in the set of data.

The mean is 82 points.

The **median** is the middle value of an ordered set of data. If there are two middle values, the median is the mean of those two values.

76, 78, **81**, 81, 94 Put the values in order first.

The median is 81 points.

The **mode** is the value that occurs most often in a set of data.

The mode is 81 points.

The **range** is the difference between the greatest and least values in the set of data.

$94 - 76 = 18$ Use subtraction to find the range.

The range is 18 points.

Find the mean, median, mode, and range of each set of values.

1. 23, 78, 45, 22

mean: 42
median: 34
mode: no mode
range: 56

2. 102, 79, 82, 103, 79

mean: 89
median: 82
mode: 79
range: 24

3. 56, 99, 112, 112, 56

mean: 87
median: 99
mode: 56, 112
range: 56

LESSON 6-2
Challenge
Speedy Data

Match each set of data with its description.

Mammal Speeds	
Antelope	54 mi/h
Cheetah	65 mi/h
Greyhound	42 mi/h
Horse	43 mi/h

C, H, M, N

Range Descriptions	
A Range	4 mi/h
B Range	29 mi/h
C Range	23 mi/h
D Range	6 mi/h

Insect Speeds	
Bumble bee	11 mi/h
Honey bee	7 mi/h
Hornet	13 mi/h
Horsefly	9 mi/h

D, G, J, N

Mean Descriptions	
E Mean	49 mi/h
F Mean	42 mi/h
G Mean	10 mi/h
H Mean	51 mi/h

Fish Speeds	
Bluefin tuna	47 mi/h
Bonefish	40 mi/h
Sailfish	69 mi/h
Swordfish	40 mi/h

B, E, L, O

Median Descriptions	
J Median	10 mi/h
K Median	42 mi/h
L Median	43.5 mi/h
M Median	48.5 mi/h

Bird Speeds	
Crane	42 mi/h
Goose	42 mi/h
Mallard	40 mi/h
Swan	44 mi/h

A, F, K, P

Mode Descriptions	
N Mode	none
O Mode	40 mi/h
P Mode	42 mi/h
Q Mode	44 mi/h

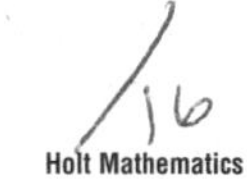

LESSON 6-2
Problem Solving
Mean, Median, Mode, and Range

Write the correct answer.

World Series Winners	
Team	**Number of Wins**
Baltimore Orioles	3
Boston Red Sox	5
Detroit Tigers	4
Minnesota Twins	3
Pittsburgh Pirates	5

1. Use the table at right to find the mean, median, mode, and range of the data set.

   mean: 4 wins; median: 4 wins;
   mode: 3 and 5 wins;
   range: 2 wins

2. When you use the data for only 2 of the teams in the table, the mean, median, and mode for the data are the same. Which teams are they?

   Orioles and Twins or Pirates and Red Sox

Circle the letter of the correct answer.

3. The states that border the Gulf of Mexico are Alabama, Florida, Louisiana, Mississippi, and Texas. What is the mean for the number of letters in those states' names?
   A 7 letters
   (B) 7.8 letters
   C 8 letters
   D 8.7 letters

4. There are 5 whole numbers in a data set. The mean of the data is 10. The median and mode are both 9. The least number in the data set is 7, and the greatest is 14. What are the numbers in the data set?
   F 7, 7, 9, 11, and 14
   G 7, 7, 9, 9, and 14
   (H) 7, 9, 9, 11, and 14
   J 7, 9, 9, 14, and 14

5. If the mean of two numbers is 2.5, what is true about the data?
   A Both numbers are greater than 5.
   B One of the numbers is less than 2.
   C One of the numbers is 2.5.
   (D) The sum of the data is not divisible by 2.

6. Tom wants to find the average height of the students in his class. Which measurement should he find?
   F the range
   (G) the mean
   H the median
   J the mode

LESSON 6-2
Reading Strategies
Vocabulary Development

The **mean**, the **median**, the **mode**, and the **range** are measures that describe a set of data.

This picture will help you learn about each measure.

Mean (average) Add all the values, then divide by the total number of values.	**Median** The middle value, when the values are arranged in order.
Mode The value that occurs most often in a set of data. There can be one mode, more than one mode, or no mode.	**Range** The difference between the greatest number and the least number in a set of data.

Measures that Describe a Set of Data

These are Tim's bowling scores from 5 different games:

89 98 110 98 105

Write **mean, median, mode,** or **range** to answer the following questions about Tim's scores.

1. The number 98 indicates which measure?

   mode or median

2. Tim added up all his scores and divided by 5. Which measure did he find?

   mean

3. Tim found that the difference between his highest and lowest score was 21 points. That measure is called the

   range

4. Tim noticed that he got a score of 98 twice. Which measure is Tim focusing on?

   mode

LESSON 6-2
Puzzles, Twisters & Teasers
Painting Mystery

For each set of numbers find the mean, the mode, and the median, in that order. Look for the letters in the boxes below and Round each answer to the nearest tenth. To solve the puzzle, read the answers(horizontally) across the mean, mode, and median columns from 1 to 12.

Numbers	Mean	Mode	Median
88, 82, 75, 75, 75	79	75	75
63, 24, 24, 37, 77, 52, 24	43	24	37
95, 100, 100, 105	100	100	100
13, 25, 47, 100, 100	57	100	47
59, 62, 62, 75, 86, 94	73	62	68.5
25, 25, 24, 24, 23, 27, 26	24.9	25 and 24	25
25, 28, 29, 26, 29, 25, 25, 30, 26	27	25	26
81, 75, 63, 51, 41, 24, 22, 18, 22	44.1	22	41
25, 30, 35, 30, 30	30	30	30
5, 15, 10, 23, 23	15.2	23	15
26, 57, 57, 33, 37	42	57	37
24, 36, 80, 24, 24	37.6	24	24

Mean Box
A = 27
M = 43
A = 57
N = 44.1
E = 24.9
O = 79
E = 100
T = 42
L = 15.2
Y = 37.6
L = 73
L = 30

Mode Box
A = 23
N = 75
C = 100
O = 24
D = 22
T = 25
E = 57
T = 24, 25
I = 62
L = 30

Median Box
A = 26
P = 30
E = 75
R = 37
H = 25
R = 100
I = 41
S = 15
K = 47
U = 24
K = 68.5

What did the painter say to the wall?

O N E M O R E C R A C K

L I K E T H A T A N D

I'L L P L A S T E R Y O U!

LESSON 6-3

Practice A

Additional Data and Outliers

Use the graph to answer Exercises 1–2.

Planet Moons

Planet	Moons
Earth	●
Mars	●●
Neptune	●●●●●●●●
Pluto	●

1. The graph shows how many moons several planets have. Find the mean, median, and mode of the data.

   mean: 3; median: 1.5; mode: 1

2. With 18 moons, Saturn has more moons than any other planet in our solar system. Add this number to the data in the graph, and find the mean, median, and mode.

   mean: 6; median: 2; mode: 1

Use the table to answer Exercises 3–4.

Astronauts in Space

Country	Number of Astronauts
France	7
Germany	9
Kazakhstan	2

3. The table shows several of the countries that have sent the most astronauts into space. Find the mean, median, and mode of the data.

   mean: 6; median: 7; mode: none

4. The United States has sent 234 astronauts into space—more than any other country. Add this number to the data in the table and find the mean, median, and mode.

   mean: 63; median: 8; mode: none

5. In Exercise 2, which measurement was affected the least by the addition of Saturn's data?

   the mode

6. Is the data in Exercise 4 best described by the mean, median, or mode?

   the median

LESSON 6-3

Practice B

Additional Data and Outliers

Use the table to answer Exercises 1–2.

Population Densities

State	People (per mi²)
Idaho	16
Nevada	18
New Mexico	15
North Dakota	9
South Dakota	10

1. The table shows population data for some of the least-crowded states. Find the mean, median, and mode of the data.

   mean: 13.6; median: 15; mode: none

2. Alaska has the lowest population density of any state. Only about 1 person per square mile lives there. Add this number to the data in the table and find the mean, median, and mode.

   mean: 11.5; median: 12.5; mode: none

Use the table to answer Exercises 3–4.

State Counties

State	Number of Counties
Illinois	102
Iowa	99
North Carolina	100
Tennessee	95
Virginia	95

3. The table shows some of the states with the most counties. Find the mean, median, and mode of the data.

   mean: 98.2; median: 99; mode: 95

4. With 254 counties, Texas has more counties than any other state. Add this number to the data in the table and find the mean, median, and mode.

   mean: $124.1\overline{6}$; median: 99.5; mode: 95

5. In Exercise 1, which measurement best describes the data? Why is Alaska's population density an outlier for that data set?

   the mean; because it is much lower than the other data

6. In Exercise 4, why is the number of counties in Texas an outlier for the data set? Which measurement best describes the data set with Texas included?

   because Texas has many more counties; the median

LESSON 6-3

Practice C

Additional Data and Outliers

Use the table to answer Exercises 1–2.

Super Bowl Winning Margins

Year	Points Won By
1967	25
1972	21
1985	22
1995	23
2001	27

1. The table shows some of the years in which the Super Bowl was won by the most points. Find the mean, median, and mode.

   mean: 23.6; median: 23; mode: none

2. The 1990 Super Bowl had the largest winning margin, which was 45 points. Add this number to the data in the table and find the mean, median, and mode. Which best describes the data?

   mean: $27.1\overline{6}$; median: 24; mode: none; median

Use the table to answer Exercises 3–4.

Successful NFL Coaches

Coach	Games Won
Paul Brown	170
Bud Grant	168
Chuck Knox	193
Chuck Noll	209
Dan Reeves	179

3. The table shows some of the most successful coaches in the NFL. Find the mean, median, and mode.

   mean: 183.8; median: 179; mode: none

4. With 347 games won, Don Shula is the most successful NFL coach. Add this number to the data in the table and find the mean, median, and mode. Which best describes the data?

   mean: 211; median: 186; mode: none; the median

5. When an outlier is added to a data set, which of these measurements will usually change the most: the range, mean, median, or mode?

   the range

6. If an outlier is greater than the other data in a set, how will it affect the mean of the data?

   The mean will always be greater.

LESSON 6-3

Reteach

Additional Data and Outliers

An **outlier** is a value in a set of data that is much greater or much less than the other values.

Number of Minutes Spent on Homework

Mon	Tue	Wed	Thurs	Fri
47	42	45	46	10

The outlier is 10 minutes, because it is much less than the other values in the set.

An outlier may affect the mean, median, or mode.

Data without Friday's value: mean = 45 median = 45.5 no mode
Data with Friday's value: mean = 38 median = 45 no mode

When Friday's value is included, the mean decreases by 7 minutes, the median decreases by 0.5 minutes, and the mode stays the same. The mean is most affected by the outlier because it is less than every value except for the outlier itself.

Find the mean, median, and mode for the set of data with and without the outlier.

1. 22, 25, 48, 26, 21, 27, 26, 29

   With outlier: mean: 28; median: 26; mode: 26

   Without outlier: mean: 25.1; median: 26; mode: 26

When an outlier affects the mean, median, or mode, choose a value that best describes the data.

In the example above, the median best describes the data because 45 minutes is closer to most of the data values in the set.

Find the mean, median, and mode. Then decide which best describes the set of data.

2. 16, 12, 14, 17, 81, 18, 13, 19, 14, 19

   mean: 22.3; median: 16.5; mode: 14 and 19. The median best describes the data.

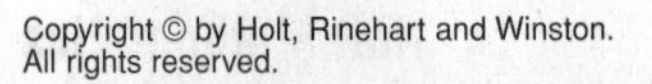

LESSON 6-3 **Challenge**

Outer Space Outlier

You have been chosen to train as an astronaut! The statisticians at NASA are not happy, because you are a major outlier for their data. Use the information below to find how you will affect their astronaut data. Some answers depend on student ages. Sample answers are given for age 12.

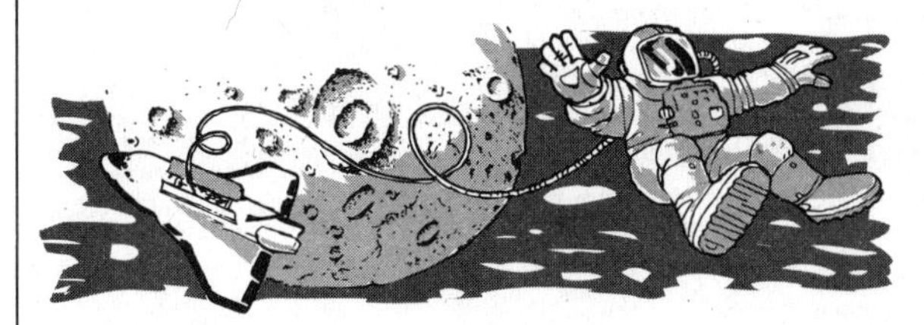

Youngest Astronauts

In 1970, Russian astronaut Gherman S. Titov became the youngest person to travel into space. He was 25 years old at liftoff. The ages of some of the other youngest astronauts of all time were 26, 29, 28, 27, 26, and 28.	
Data Without Your Age:	**Data With Your Age:**
Mean age: 27	**Mean age:** 25.125
Median age: 27	**Median age:** 26.5
Mode age: 26 and 28	**Mode age:** 26 and 28

Oldest Astronauts

In 1998, American astronaut John H. Glenn became the oldest person to travel into space. He was 77 years old at liftoff. The ages of some of the other oldest astronauts of all time were 54, 59, 61, 56, 58, and 55.	
Data Without Your Age:	**Data With Your Age:**
Mean age: 60	**Mean age:** 54
Median age: 58	**Median age:** 57
Mode age: no mode	**Mode age:** no mode

LESSON 6-3 **Problem Solving**

Additional Data and Outliers

Use the table to answer the questions.

Successful Films in the U.S.

Film	U.S. Earnings for first release (million $)
E.T. the Extra-Terrestrial	400
Forrest Gump	330
Independence Day	305
Jurassic Park	357
The Lion King	313

1. Find the mean, median, and mode of the earnings data.

   mean: $341 million; median: $330 million; mode: none

2. *Titanic* earned more money in the United States than any other film—a total of $600 million! Add this figure to the data and find the mean, median, and mode. Round your answer for the mean to the nearest whole million.

   mean: $384 million; median: $343.5 million; mode: none

Circle the letter of the correct answer.

3. In Canada, people watch TV an average of 74 minutes each day. In Germany, people watch an average of 68 minutes a day. In France it is 67 minutes a day, in Spain it is 91 minutes a day, and in Ireland it is 74 minutes a day. Find the mean, median, and mode of the data.
   - **A** mean: 74 min.; median: 74 min.; mode: 74 min.
   - **B** mean: 74 min.; median: 74.8 min.; mode: 74 min.
   - **C** mean: 74.8 min.; median: 74 min.; mode: 24 min.
   - **(D)** mean: 74.8 min.; median: 74 min.; mode: 74 min.

4. People in the United States watch more television than in any other country. Americans watch an average of 118 minutes a day! Add this number to the data and find the mean, median, and mode.
   - **(F)** mean: 82 min.; median: 74 min.; mode: 74 min.
   - **G** mean: 82 min.; median: 74 min.; mode: 118 min.
   - **H** mean: 82 min.; median: 91 min.; mode: 74 min.
   - **J** mean: 74.8 min.; median: 82 min.; mode: 74 min.

5. In Exercise 2, which data measurement changed the least with the addition of *Titanic's* earnings?
   - **A** the range
   - **(C)** the median
   - **B** the mean
   - **D** the upper extreme

6. In Exercise 4, which measurements best describe the data?
   - **F** mean and median
   - **G** range and mean
   - **(H)** median and mode
   - **J** range and mode

LESSON 6-3 **Reading Strategies**

Use Graphic Aids

Tim put his bowling scores on a number line.

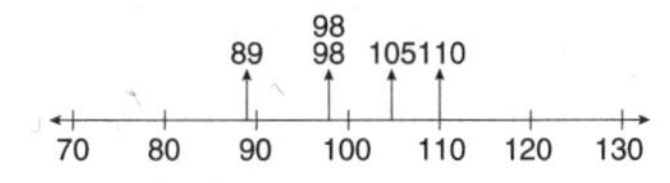

The number line lets you see whether the scores are close together or spread apart.

Recall the measures that describe Tim's scores:

Mean—100 Median—98 Mode—98 Range—21

Tim bowled another game and got a score of 70. This number line includes Tim's new score.

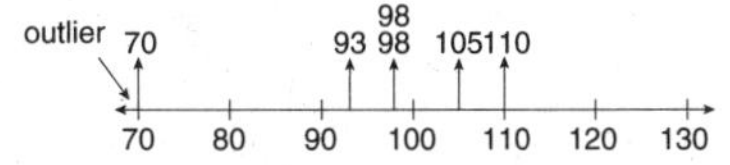

The score of 70 is called an **outlier,** because it is set apart from the other scores.

Answer the following questions.

1. How does the number line help you see the outlier in these scores?

   Possible answer: You can see that 70 is set apart from the other scores.

2. Will the mean increase or decrease when the score of 70 is included?

   decrease

3. How does the number line help you find the mode?

   Possible answer: The numbers are stacked on top of each other.

4. With the addition of the score of 70, will the range increase or decrease?

   increase

5. Circle the correct answer: Which measure is not changed with the added score of 70?

   median (mode)

LESSON 6-3 **Puzzles, Twisters & Teasers**

A-Maze-ing Data!

First, answer each question. Then use your answers to navigate through the maze.

1. Consider the data set 2, 4, 6, 8, 8, 8, 10, 12, 14, 16. What happens to the mean if you add the values 3 and 9?

   The mean goes down by 0.5 . (5 tenths)

2. What is the mean if you add 2 and 18 to the original data set? 9

3. As CEO of a company you notice that your five executives have the following salaries: $65,000, $70,000, $80,000, $80,000, and $72,000. You are hiring a sixth executive and will pay her $78,000. By how many thousands did the median change? 3

4. Consider the data set 6, 8, 10, 12, 12, 12, 14, 16, 18, 20. What happens to the mean if you add the values 22 and 22.7? The mean goes up by 1.6 .

5. Consider the data set: 6, 8, 12, 13 and 16. Which statistical measure will change if you add 8 and 14 to the data set?
   - If Mean move 1 space right
   - If Median move 4 spaces left
   - (If Mode move 5 spaces down)
   - If none move 3 spaces right

Now you must find your way through this maze to increase the mean of your grades.

Go to start and move down the amount of answer #1 times 10.

Move right the amount of answer #2.

Move up the amount of answer #3.

Move right the amount of answer #4 rounded to the nearest whole number.

Follow the directions for the answer you selected for #5.

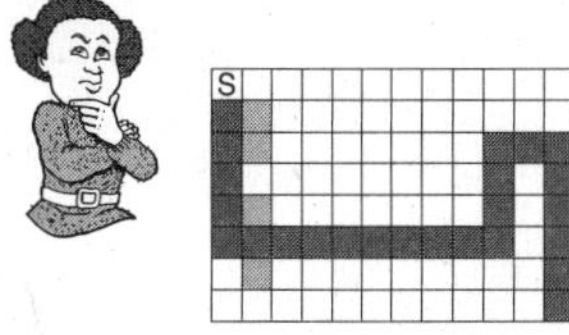

LESSON 6-4
Practice A
Bar Graphs

Use the bar graph to answer each question.

1. What is the most common city name in the United States?

   Fairview

2. Which two names on the graph are used for the same number of cities?

   Franklin and Riverside

3. Which name is used for 52 different cities in the United States?

   Midway

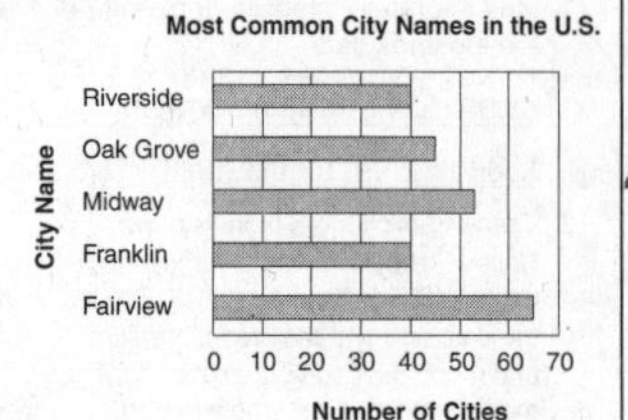

Use the given data to make a bar graph.

Skyscrapers in Some Cities

City, State	Skyscrapers
Chicago, IL	75
Dallas, TX	20
Houston, TX	30
Los Angeles, CA	22
Hong Kong, China	42
Tokyo, Japan	30

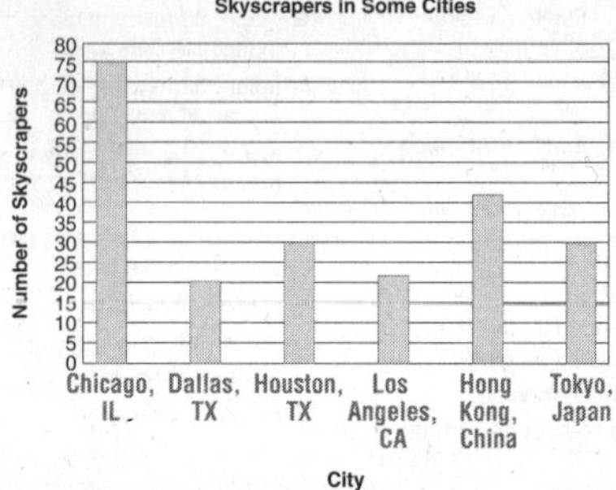

LESSON 6-4
Practice B
Bar Graphs

Use the bar graph to answer each question.

1. In which country did people spend the most money on toys in 2000?

   the United States

2. In which two countries did people spend the same amount of money on toys in 2000? How much did they each spend?

   France and Germany; $3 million each

3. In which country did people spend $9 million on toys in 2000?

   Japan

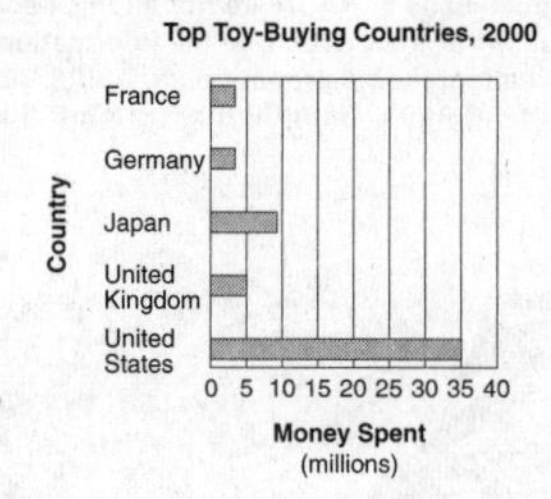

Make a bar graph to compare the data in the table.

Female Groups with the Most Top 10 and Top 20 Hits

Top 10		Top 20	
The Supremes	20	The Supremes	24
The Pointer Sisters	7	The Pointer Sisters	13
TLC	9	TLC	11
En Vogue	5	En Vogue	7
Spice Girls	4	Spice Girls	7

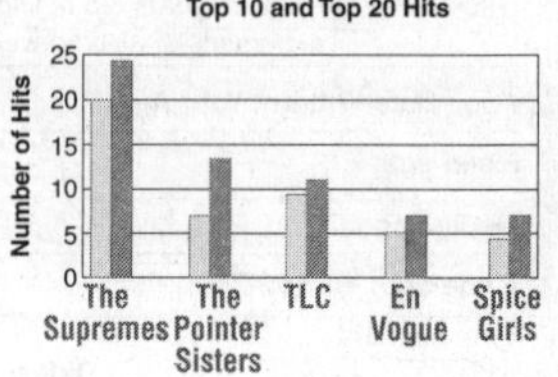

LESSON 6-4
Practice C
Bar Graphs

Use the bar graph to answer each question.

1. How did the number of Asian immigrants coming to the United States compare for 1900 and 2000?

   More immigrants came in 2000.

2. From which region did most people immigrate in 1900?

   Europe

3. From which region did most people immigrate in 2000?

   Latin America

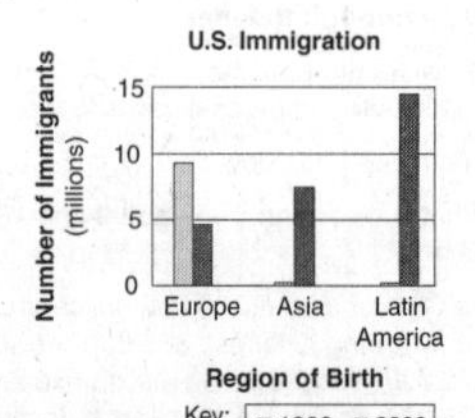

Use the given data to make a bar graph and answer the questions.

U.S. Foreign-born Population

Year	Males	Females
1900	5,630,190	4,711,086
1950	5,258,255	5,089,140
2000	14,200,000	14,179,000

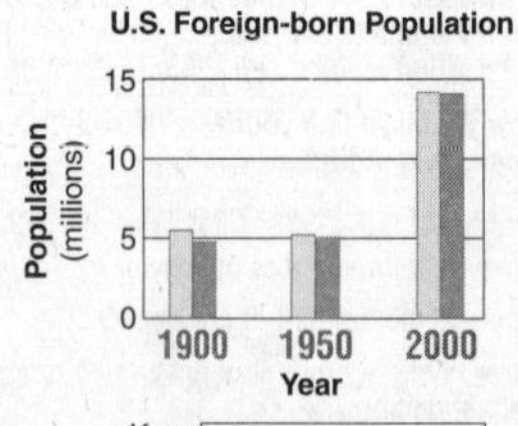
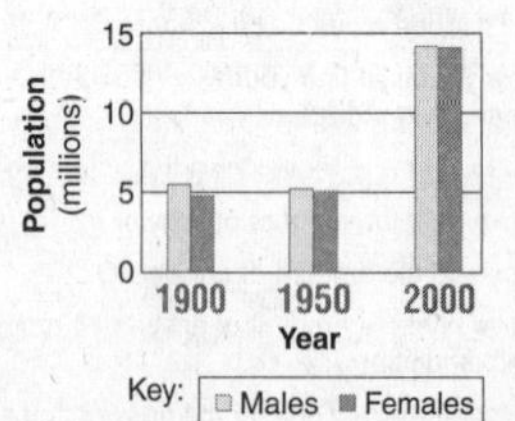

4. Which year had the highest total foreign-born population?

   2000

5. How did the male and female foreign-born populations compare between 1900 and 2000?

   The male foreign-born population was higher.

LESSON 6-4
Reteach
Bar Graphs

You can make a bar graph to compare amounts.

To make a bar graph using the data in the table, first choose a scale that includes all of the data values. Next, separate the scale into equal parts, called intervals.

Then draw bars to match the data. The bars should be of equal width and should not touch. Give your graph a title and label its axes.

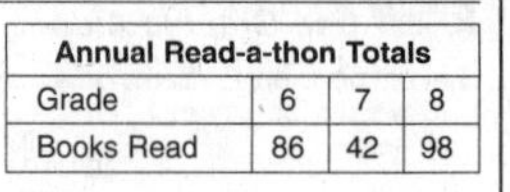

Annual Read-a-thon Totals

Grade	6	7	8
Books Read	86	42	98

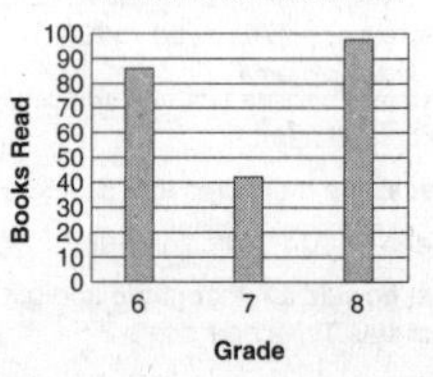

Use the data to make a bar graph.

1.

Canned Food Drive Totals

Grade	6	7	8
Cans Collected	96	74	62

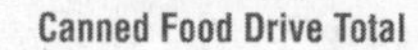

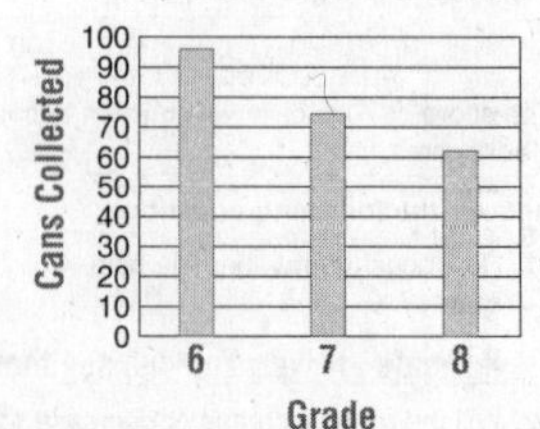

Look at the bar graph for the Read-a-thon above. Which grade read almost twice as many books as the seventh grade?

The bar for the sixth grade is about twice a long as the bar for the seventh grade. So the sixth grade read almost twice as many books as the seventh grade.

Use the bar graph you made in Exercise 1.

2. How many more items did the sixth grade collect than the eighth grade?

   The sixth grade collected 34 more items.

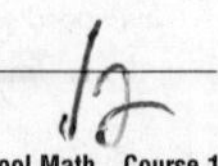

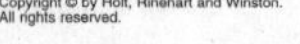

LESSON 6-4
Reteach
Bar Graphs (continued)

A double-bar graph shows two sets of related data.

To make a double-bar graph, choose a scale and an interval for the scale. Then draw bars to match the data. The bars for the same grade should touch, but bars for different grades should not touch.

Middle School Students			
Grade	6	7	8
Boys	34	29	25
Girls	28	31	22

Because there are two bars for each grade, make a key to show which bars represent girls and which bars represent boys.

Your graph should have a title and its axes should be labeled.

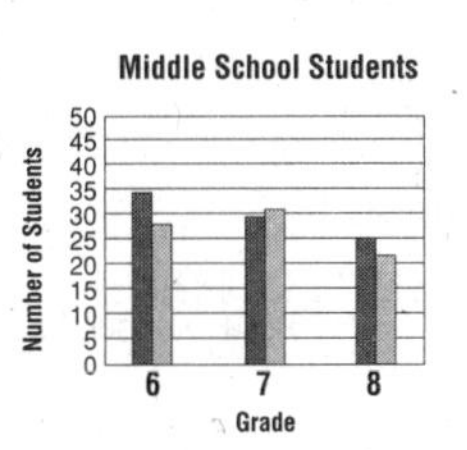

Make a double-bar graph to match the data below. Then answer the question.

3.

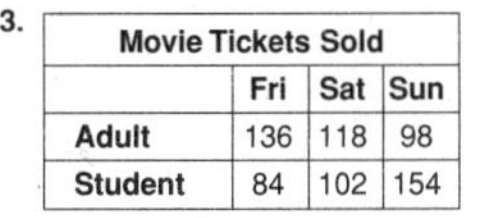

Movie Tickets Sold			
	Fri	Sat	Sun
Adult	136	118	98
Student	84	102	154

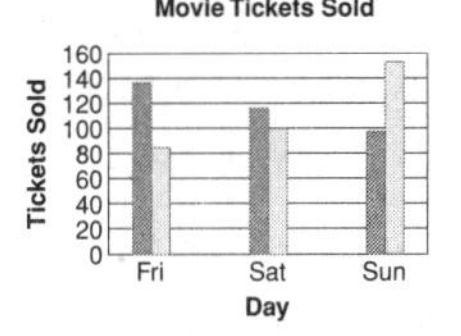

4. On which day was the number of adult tickets sold about the same as the number of student tickets sold?

The number of adult tickets sold was about the same as the number of student tickets sold on Saturday.

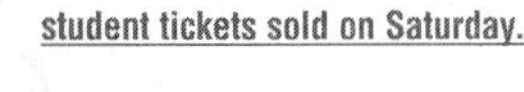

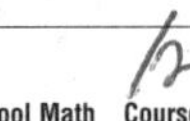

LESSON 6-4
Challenge
Picture a Bar Graph

Sometimes people use illustrations instead of bars to display data on bar graphs. For example, a bar graph showing the sizes of some forests might use trees for the graph's bars.

Use the data given in each table below to make an illustrated bar graph. Make sure the pictures you choose for your bars relate to the subject of each graph.

Fastest U.S. Roller Coasters

Roller Coaster, Location	Speed (mi/h)
Desperado, NV	80
Goliath, CA	85
Millennium Force, OH	92
Superman: The Escape, CA	100
Titan, TX	85

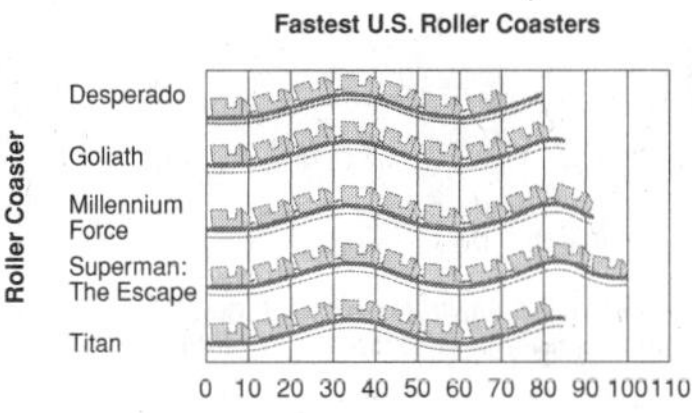

Check students' graphs. Ask students to explain why they chose their illustrations for each graph's bars.

Surfers with the Most World Championship Wins

Surfer, Country	Wins
Tom Carroll, Australia	2
Tom Curren, United States	3
Damien Hardman, Australia	2
Mark Richards, Australia	4
Kelly Slater, United States	6

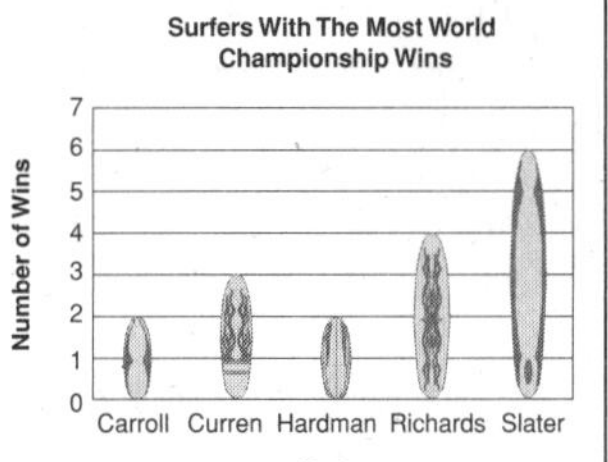

LESSON 6-4
Problem Solving
Bar Graphs

Use the bar graph for Exercises 1–4.

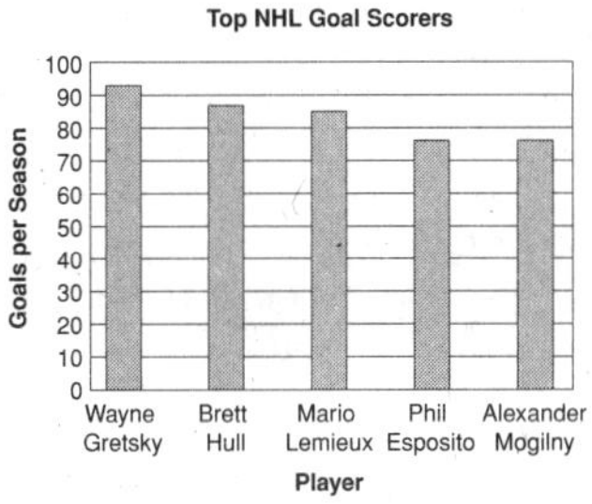

Top NHL Goal Scorers (Goals per Season, 0–100; Player: Wayne Gretsky, Brett Hull, Mario Lemieux, Phil Esposito, Alexander Mogilny)

1. What is the range of the goals the hockey players scored per season?
16 goals
2. What is the mode of the goals scored?
76 goals
3. What is the mean number of goals the players scored?
83 goals

Use the bar graph for Exercises 5–8.

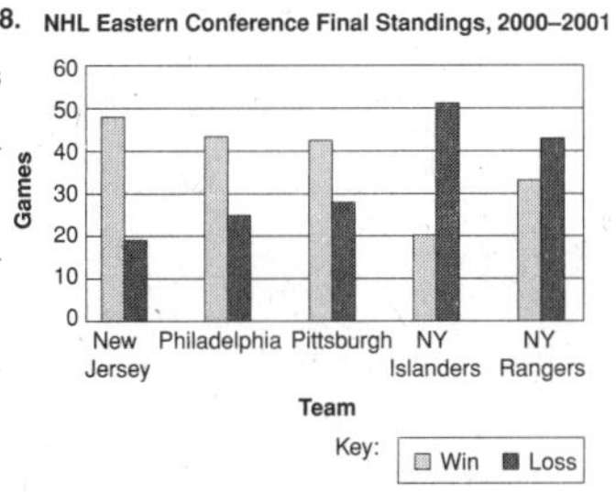

NHL Eastern Conference Final Standings, 2000–2001 (Games, 0–60; Team: New Jersey, Philadelphia, Pittsburgh, NY Islanders, NY Rangers; Key: Win, Loss)

4. Which team won the most games that season? New Jersey
5. Which team lost the most games that season? NY Islanders
6. What was the mean number of games won? 37 games
7. What was the mean number of games lost? 33 games

Circle the letter of the correct answer.

8. Which hockey team had the greatest difference between the number of games won and lost?
A New Jersey
(B) New York Islanders
C Philadelphia
D Pittsburgh

9. How do you know the mode of a data set by looking at a bar graph?
(F) The mode depends on the type of data displayed in the bar graph.
G The mode is the first bar.
H The mode has the lowest bar.
J The bar for the mode is in the middle of the graph.

LESSON 6-4
Reading Strategies
Compare and Contrast

A **bar graph** shows data and makes it possible to compare facts about that data. A group of sixth-grade students voted on their favorite kind of exercise. The graph shows the data that was collected from the votes.

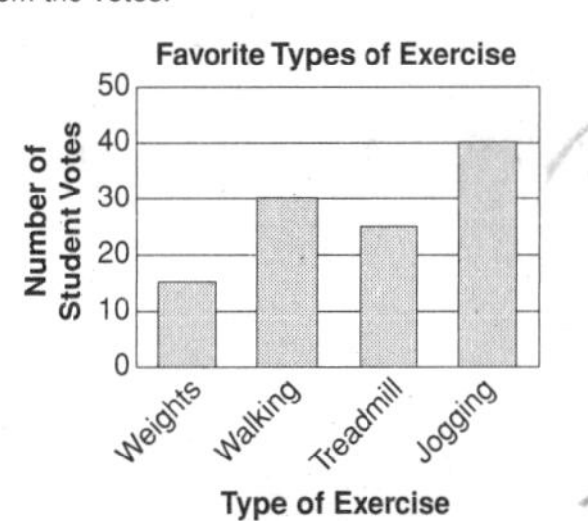

Answer the following questions about the bar graph.

1. What does the graph show? favorite types of exercises
2. What does the scale of the graph count by? tens
3. How can you tell from the graph which exercise had the most votes?
Possible answer: The tallest bar stands for the most votes.
4. Compare votes for walking and votes for jogging. Which one got the most votes? jogging
5. Compare votes for the treadmill and votes for weights. Which one had the most votes? treadmill
6. Which exercise had the least amount of votes on the graph? weights
7. How can you tell from looking at the graph which exercise had the fewest votes?
Possible answer: The shortest bar stands for the fewest votes.
8. How do the bars on the graph help you compare data?
Possible answer: The length of each bar helps you compare the data.

LESSON 6-4
Puzzles, Twisters & Teasers
At The Movies

Ten students were surveyed to see how many times they went to the movies last month.

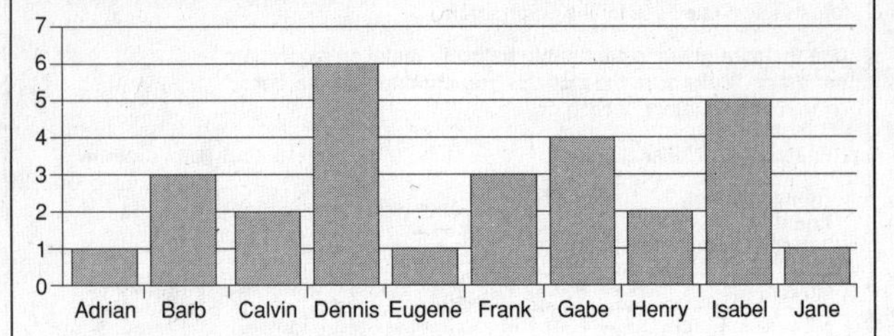

To solve the riddle answer each question. Use the letter in front of each answer to fill in the blanks of the riddle.

1. How many students went to the movies more than four times last month?
   S = 1   (T)= 2   L = 3   R = 4
2. Who went to the movies the most last month?
   I = Calvin   E = Gabe   (A)= Dennis   O = Isabel
3. How many students went to the movies less than three times last month?
   R = 3   S = 4   (B)= 5   C = 6
4. How many students went to the movies only once last month?
   W = 0   L = 1   K = 2   (U)= 3
5. How many more times did Dennis go to the movies than Gabe?
   (K)= 2   N = 3   G = 4   D = 5
6. Frank and Barb went to the movies a total of how many times?
   L = 3   T = 4   A = 5   (C)= 6
7. Who went to twice as many movies as Calvin?
   Y = Barb   N = Frank   (O)= Gabe   D = Isabel

Where do Australian children go to play?

O	U	T	B	A	C	K
7	4	1	3	2	6	5

/8

LESSON 6-5
Practice A
Line Plots, Frequency Tables, and Histograms

5 1. Hockey players voted for a team name. The results are shown in the box. Make a tally table. Which name got the fewest votes?

Bulldogs

Bears	Wildcats	Bulldogs	Lions	Bears	Wildcats
Bears	Bears	Wildcats	Bears	Lions	Bears

Tally Table for Hockey Team Name Votes

Bears	𝍸 I
Bulldogs	I
Lions	II
Wildcats	III

2. Make a line plot of the data.

Number of Goals Scored by 25 Hockey Players										
0	2	4	2	1	0	2	5	3	2	1	3
0	1	4	3	1	2	3	1	4	5	1	2

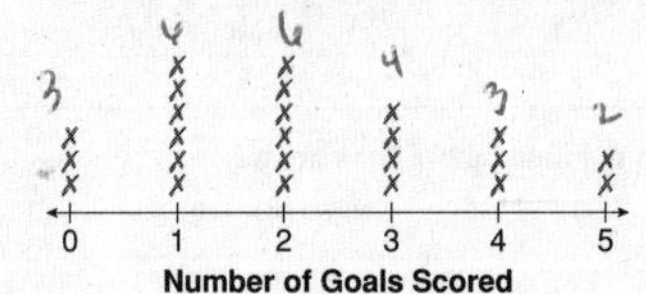

6 3. Use the data in the box below to complete the frequency table with intervals.

Ages of Hockey Fans Polled at Tonight's Game									
14	10	38	54	27	29	7	16	10	45
18	21	9	36	25	17	39	33	26	30

Ages of Hockey Fans Polled at Tonight's Game

Ages	1–10	11–20	21–30	31–40	41–50	51–60
Frequency	4	4	6	4	1	1

4. To which age group did the most fans belong? **21–30**

LESSON 6-5
Practice B
Line Plots, Frequency Tables, and Histograms

1. Students voted for a day not to have homework. The results are shown in the box. Make a tally table. Which day got the most votes?

Friday

Monday	Friday	Thursday	Friday	Tuesday	Friday
Friday	Thursday	Wednesday	Monday	Friday	Monday

Tally Table for Homework Votes

Mon	III
Tues	I
Wed	I
Thurs	II
Fri	𝍸

2. Make a line plot of the data.

Average Time Spent on Homework Per Day (min)									
20	21	24	20	21	20	20	22	25	20
22	20	24	25	24	25	25	21	25	24

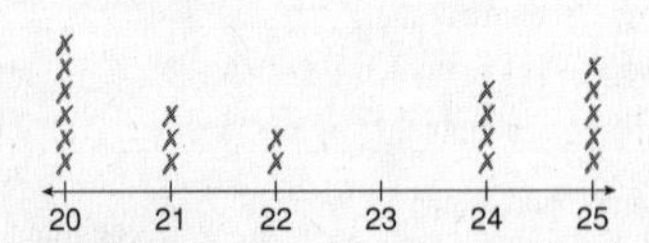

3. Use the data in the box below to make a frequency table with intervals.

Class Social Studies Test Scores									
78	95	81	83	75	68	100	73	92	85
59	70	88	92	99	87	75	67	89	84

Class Social Studies Test Scores

Scores	51–60	61–70	71–80	81–90	91–100
Frequency	1	3	4	7	5

Possible answer:

4. In which range of scores did most of the students' tests fall? **81-90**

LESSON 6-5
Practice C
Line Plots, Frequency Tables, and Histograms

5 1. During a car trip, Ed counted the number of red, black, blue, and green cars he passed on the highway. The results are shown in the box. Make a tally table. Which color of car did Ed count most often?

Red

red	black	blue	green	red	blue	red	black
blue	green	red	black	red	red	black	

Tally Table for Car Colors

Red	𝍸 I
Black	IIII
Blue	III
Green	II

2. Make a line plot of the data.

Ages of Drivers at Truck Stop									
17	28	29	34	47	55	27	18	39	42	22
33	21	46	52	55	41	48	19	23	25	

16 17 18 19 20 21 22 23 24 25 26 27 28 29 30 31 32 33 34 35 36 37 38 39 40 41 42 43 44 45 46 47 48 49 50 51 52 53 54 55

Ages of Drivers at a Truck Stop

3. Use the data in the box above to make a frequency table with intervals.
4. Use your frequency table from Exercise 3 to make a histogram.
5. To which age group did most of the drivers belong? **16–25**

Ages of Drivers at Truck Stop

Age Intervals	Frequency
16–25	7
26–35	5
36–45	3
46–55	6

Possible answers:

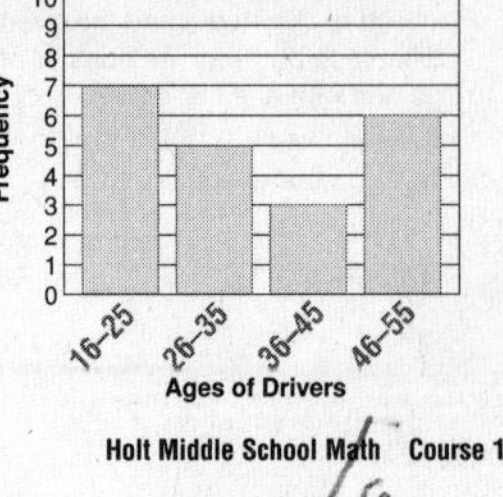

/9

LESSON 6-5 **Reteach**

Line Plots, Frequency Tables, and Histograms

Julie picked the following cards from a deck.

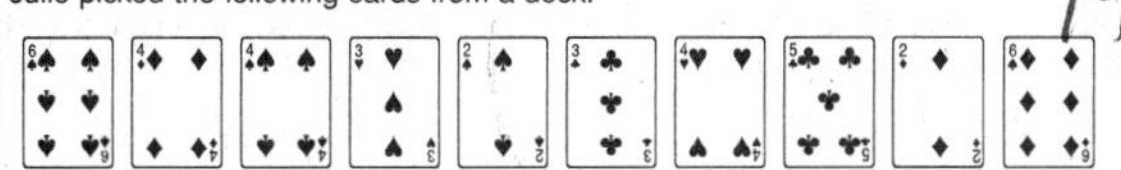

You can make a tally table to organize the data. Make a row for the numbers. Then for each card, make a tally mark in the appropriate column.

Julie's Cards

2	3	4	5	6
II	II	III	I	II

1. Make a tally table to organize the data.

Rolls of a Number Cube						
2	3	6	5	1	4	1
3	3	5	1	6	1	4

Rolls of a Number Cube					
1	2	3	4	5	6
IIII	I	III	II	II	II

A line plot gives a visual picture of data. To make a line plot of Julie's data, draw a number line. Then use an X to represent each tally mark in the tally table.

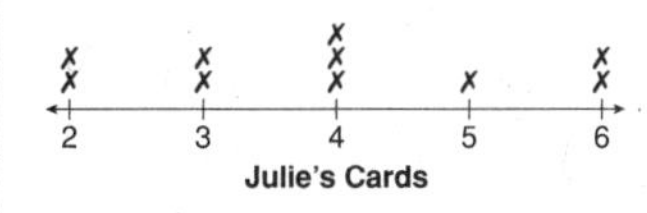

2. Make a line plot for the tally table you made Exercise 1.

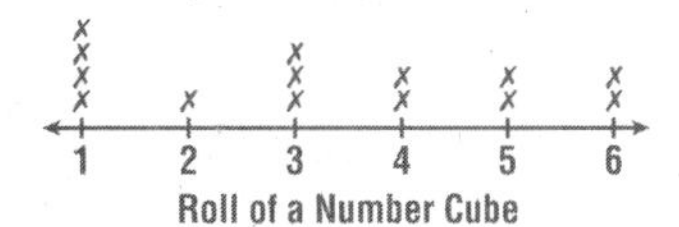

LESSON 6-5 **Reteach**

Line Plots, Frequency Tables, and Histograms

Sometimes, you can make a frequency table with intervals or a histogram.

Number of Jumping Jacks Completed in 30 Seconds				
12	28	24	32	35
31	38	55	43	52
42	49	18	22	15
47	37	19	31	37

A frequency table can organize the data with intervals.

Jumping Jacks

Interval	Frequency
1–10	0
11–20	4
21–30	3
31–40	7
41–50	4
51–60	2

A histogram is a bar graph that shows the number of values that occur within each interval.

You make a histogram the same way you make any other bar graph, except that the bars touch. They do not overlap.

Here is a histogram for the frequency table above.

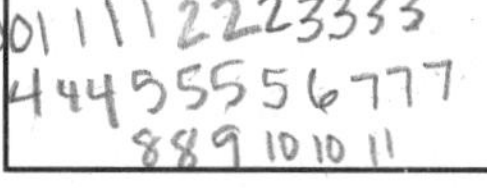

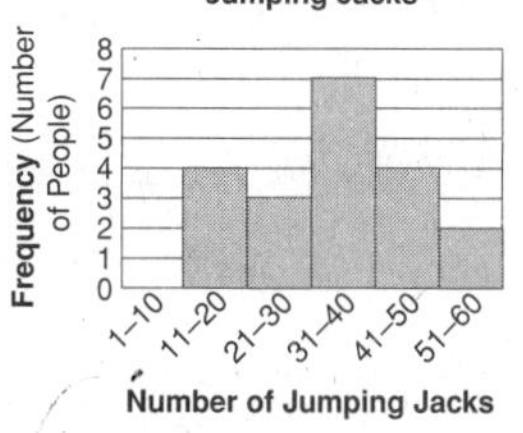

3. Use the data to make a histogram.

Total Books Read by Participants in Summer Reading Program				
5	3	8	7	6
2	9	10	1	2
4	5	7	3	5
3	1	0	10	4
3	5	8	2	1
1	7	0	4	11

Possible answer:

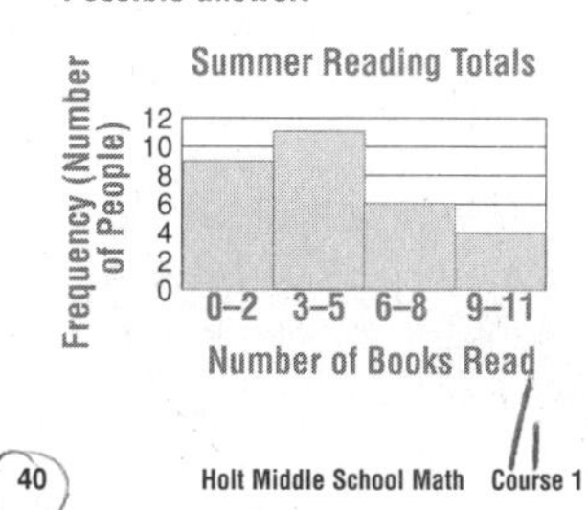

LESSON 6-5 **Challenge**

Write Often

What letter is used more than any other letter in the English language? e

The box below contains the six English letters that are used most often. Use the box to complete the tally table at the bottom of the page. Your completed table will show the answer to the question.

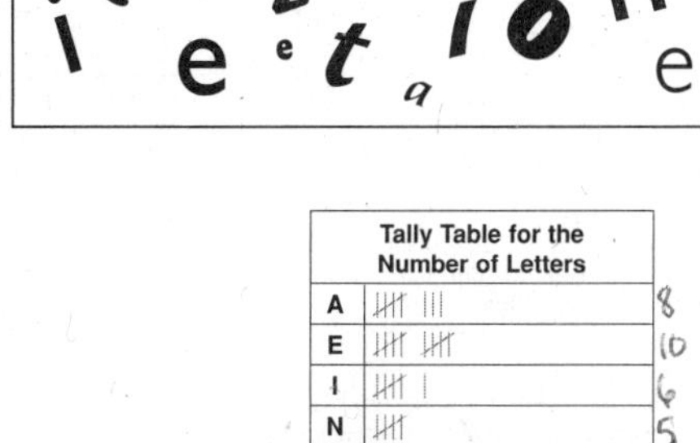

Tally Table for the Number of Letters	
A	卌 III
E	卌 卌
I	卌 I
N	卌
O	卌 II
T	卌 IIII

LESSON 6-5 **Problem Solving**

Line Plots, Frequency Tables, and Histograms

The sixth grade class voted on their favorite ice cream flavors. The results of the vote are shown below.

chocolate	vanilla	strawberry	vanilla	vanilla
vanilla	chocolate	vanilla	chocolate	strawberry
chocolate	strawberry	vanilla	vanilla	chocolate

1. Use the data to make a tally table. How many students voted in all?
   15 students
2. Which flavor got the most votes?
   vanilla

Ice Cream Flavor Votes

Flavor	Number of Votes
Chocolate	卌
Vanilla	卌 II
Strawberry	III

Use the histogram for Exercises 3–5.

3. How many years make up each age interval on the histogram?
   20 years
4. Which range of ages on the histogram has the highest population?
   25–44
5. Which range of ages has the lowest population?
   65–84

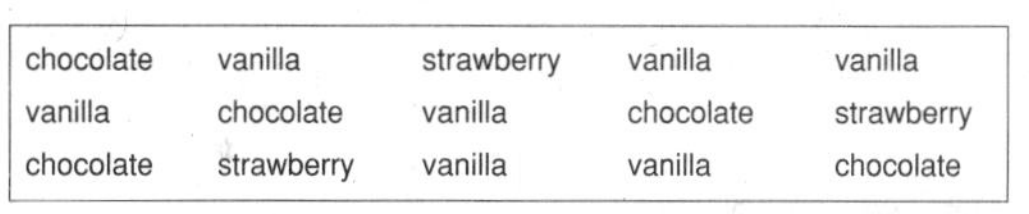

Circle the letter of the correct answer.

6. Which of the following cannot be used to make a frequency table with intervals?
   A histogram
   B tally table
   C line plot
   (D) double-bar graph
7. Which question can be answered by using the histogram above?
   A How many people in the United States are younger than 5 years?
   B What is the mean age of all people in the United States?
   C How many people in the United States are older than 84 years old?
   (D) How many people in the United States are age 25 to 64?

LESSON **6-5**

Reading Strategies
Reading a Table

This is a **tally table** showing the number of cartons of milk sold to sixth-graders in one week.

Day	Number of Milk Cartons
Monday	卌 卌 卌 II
Tuesday	卌 卌 卌
Wednesday	卌 卌 II
Thursday	卌 卌 III
Friday	卌 卌 IIII

Answer each question about the tally table.

1. How are the tally marks organized? in groups of 5
2. For how many days are milk sales shown? 5 days
3. How can you tell by looking at the table which day had the most sales?

   Possible answer: The longest row of tally marks shows the day with the most sales.

Creating a **frequency table** is one way to summarize the data in the tally chart. This table organizes data by how often it occurs.

Day	Frequency (Number of cartons sold)
Monday	17
Tuesday	15
Wednesday	12
Thursday	13
Friday	14

Use the frequency table to answer the following questions.

4. How is the data in the tally table shown differently than data in the frequency table?

   Numbers are used instead of tally marks.

5. How many cartons of milk were sold on Tuesday? 15 cartons
6. After lunch on Wednesday, how many cartons of milk had been sold so far that week? 44 cartons

LESSON **6-5**

Puzzles, Twisters & Teasers
Bertha's Burger Bonanza

Bertha's menu includes hamburgers, cheeseburgers, hot dogs, chilidogs, French fries, milk shakes, chicken, juice, and tacos. One day, to speed up things, Bertha herself wrote down everything the people in line wanted that day. Here is her list.

Chili Dog	Juice	Tacos	Cheeseburger	Chicken
Hamburger	Juice	Cheeseburger	Juice	Hamburger
Hot Dog	French Fries	Juice	Hot Dog	Chicken
Milk Shake	Chili Dog	Juice	Juice	Chicken
French Fries	Hamburger	Tacos	French Fries	Chili Dog
Hamburger	Chicken	Cheeseburger	Cheeseburger	Hamburger
Tacos	Hamburger	Chicken	Tacos	Hamburger
French Fries	Chicken	French Fries	Tacos	Juice
Hamburger	Tacos	Tacos	Tacos	Tacos

Complete the table below using the data from Bertha's list.

Number of Foods Ordered at Bertha's

Type of Food	Frequency	
Hamburger	8	U
Cheeseburger	4	T
Hot Dog	2	A
Chili Dog	3	H
French Fries	5	N
Chicken	6	I
Milk Shake	1	E
Tacos	9	P
Juice	7	R

To solve the riddle, find the letter that corresponds to each frequency. Rearrange the letters to solve the riddle. ** The letter for chicken is used twice. **

8	18	1	32	7	9	19	5	16	23	6	30	4	7	13	2	28	22	11	15	3	18
U	D	E	K	R	P	B	N	W	G	I	Z	T	M	F	A	S	C	J	Y	H	O

The skydiver had fries and a milk shake. Why was he nervous?

Because he was U P I N T H E A I R!

LESSON **6-6**

Practice A
Ordered Pairs

Use the coordinate grid to answer Exercises 1–2. Circle the letter of the correct answer.

1. Which ordered pair gives the location of the roller coaster in the park?
   - A (5, 3)
   - B (3, 3)
   - (C) (3, 5)
   - D (5, 5)
2. Which feature on the coordinate grid of the park is located at point (1, 2)?
   - (F) Ferris wheel
   - G Roller coaster
   - H Snack bar
   - J Water slide

Use the coordinate grid to answer Exercises 3–4. Circle the letter of the correct answer.

3. Which ordered pair describes the location of point *C*?
   - A (1, 3)
   - B (3, 1)
   - (C) $(3\frac{1}{2}, 0)$
   - D (0, 3)
4. Which point is located at (1, 3) on the grid?
   - F point A
   - G point B
   - H point C
   - (J) point *D*
5. The restrooms at the amusement park are located at point (4, 4). Label the restrooms on the grid for Exercises 1–2. Check students' graphs.
6. Point *E* is located 1 unit up and 2 units to the right of point *A* on the grid above. Label point *E* on the grid for Exercises 3–4. Check students' graphs for *E*(5, 2).

LESSON **6-6**

Practice B
Ordered Pairs

Name the ordered pair for each location on the grid.

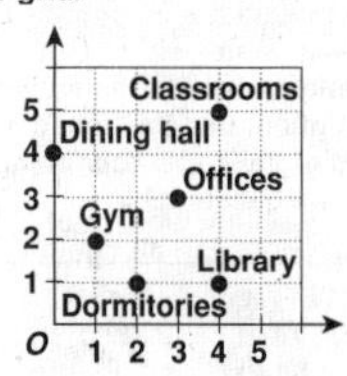

1. gym (1, 2)
2. dining hall (0, 4)
3. offices (3, 3)
4. library (4, 1)
5. classrooms (4, 5)
6. dormitories (2, 1)

Graph and label each point on the coordinate grid.

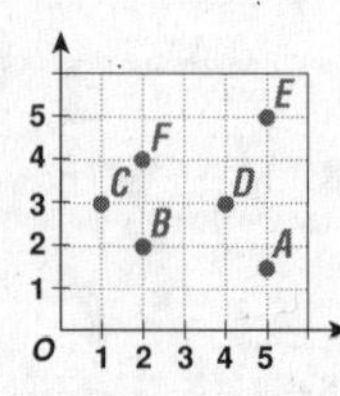

7. $A(5, 1\frac{1}{2})$
8. *B* (2, 2)
9. *C* (1, 3)
10. *D* (4, 3)
11. *E* (5, 5)
12. *F* (2, 4)
13. On a map of his neighborhood, Mark's house is located at point (7, 3). His best friend, Cheryl, lives 2 units west and 1 unit south of him. What ordered pair describes the location of Cheryl's house on their neighborhood map?

    (5, 2)

14. Quan used a coordinate grid map of the zoo during his visit. Starting at (0, 0), he walked 3 units up and 4 units to the right to reach the tiger cages. Then he walked 1 unit down and 1 unit left to see the pandas. Describe the directions Quan should walk to get back to his starting point.

    walk 2 units down and 3 units to the left

LESSON 6-6 **Practice C**
Ordered Pairs

Give the ordered pair for each location.

1. firehouse (6, 8)
2. police station (1, 7)
3. city hall (3, 1)
4. zoo $(1, 2\frac{1}{2})$
5. park (7, 5)
6. post office (9, 3)

Name the point found at each location.

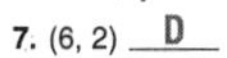

7. (6, 2) D
8. (9, 2) E
9. (9, 5) F
10. (6, 5) G
11. (1, 6) A
12. $(3, 8\frac{1}{2})$ B
13. (5, 6) C

14. Draw lines to connect the points *A* and *B*, *B* and *C*, and *C* and *A*. What shape is formed?

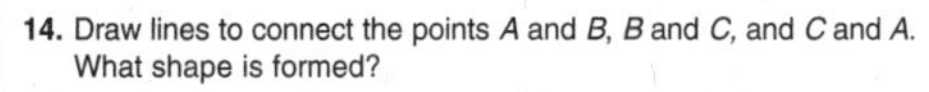

a triangle

15. Draw lines to connect the points *D* and *E*, *E* and *F*, *F* and *G*, and *G* and *D*. What shape is formed?

a square

16. If you wanted to create a square using points *A* and *C* on the grid above, how many new points would you have to draw? What ordered pairs describe those new points?

2 new points; Possible answer: (1, 10) and (5, 10)

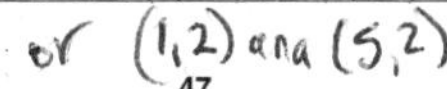

or (1,2) and (5,2)

18

LESSON 6-6 **Reteach**
Ordered Pairs

A coordinate grid is formed by horizontal and vertical lines and is used to locate points.

An ordered pair names the location of a point by using two numbers.

The ordered pair (2, 5) gives the location of point *A* on the coordinate grid.

The first number, 2, tells the horizontal distance from the starting point (0, 0).

The second number, 5, tells the vertical distance.

To find the ordered pair for point *B*, start at (0, 0). Then move 6 units right and $3\frac{1}{2}$ units up. The coordinates of point *B* are $(6, 3\frac{1}{2})$.

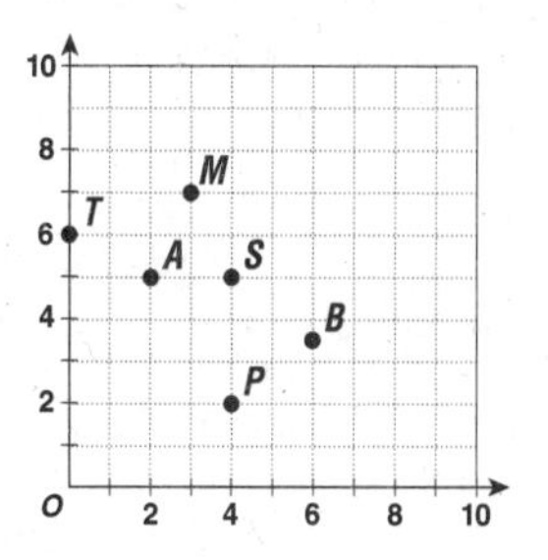

Give the ordered pair for each point shown on the coordinate grid above.

1. *P* (4, 2)
2. *T* (0, 6)
3. *M* (3, 7)
4. *S* (4, 5)

You can plot points in a coordinate grid.

To plot *F* (6, 4), start at (0, 0). Then move 6 units right and 4 units up.

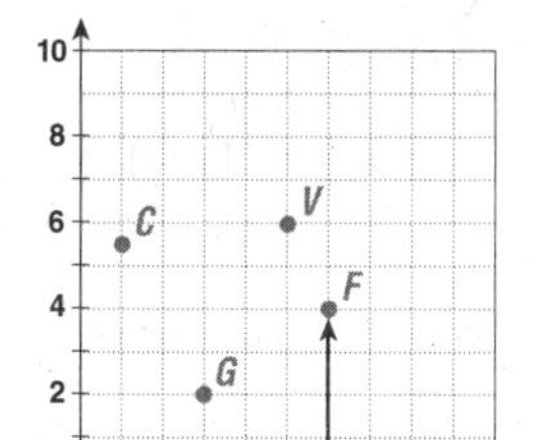

Plot each point in the coordinate grid above.

5. *V* (5, 6)
6. *G* (3, 2)
7. *K* (7, 0)
8. $C (1, 5\frac{1}{2})$

18

LESSON 6-6 **Challenge**
Treasure Island

According to legend, a pirate named Blackbeard buried his stolen treasures somewhere on the Outer Banks, off the coast of North Carolina. While on vacation there, you found an old sea chest buried on the beach. Inside it was a map that just may lead you to Blackbeard's hidden treasures!

Follow the map's clues to each location on the island. For each clue, name the location and the ordered pair that describes its point on the map. Graph each of those points on the map. Then draw an *X* where you think the treasure is buried.

2 each

1. Land your ship 7 units east and 6 units north of the (0, 0) starting point. Pirates Cove; (7, 6)
2. Walk 5 units west. Fool's Forest; (2, 6)
3. Walk 5 units south and 1 unit east. Skipper's Village; (3, 1)
4. Walk 2 units east and 2 units north. Terror Hills; (5, 3)
5. Walk 2 units north and 1 unit east. Silver Lake; (6, 5)
6. Walk 1 unit east and 3 units north. Deadman's Swamp; (7, 8)
7. Walk 3 units west and 1 unit north to find my treasure. Shipwreck Rocks; (4, 9)

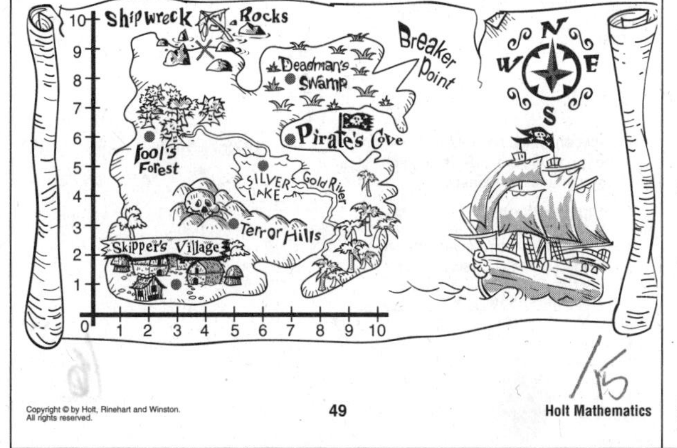

15

LESSON 6-6 **Problem Solving**
Ordered Pairs

Use the coordinate grid to answer each question.

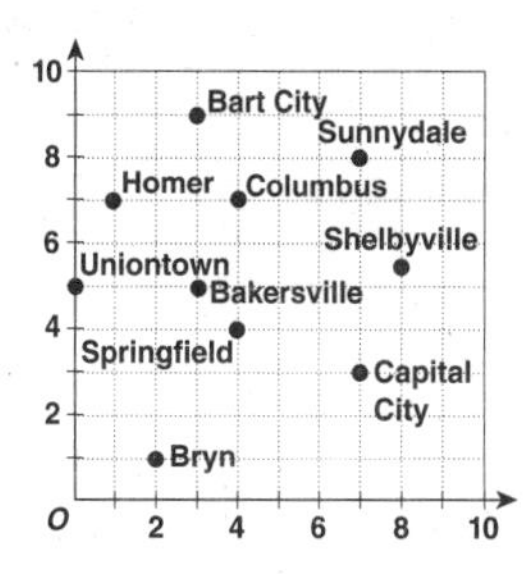

1. What city is located at point (4, 4) on the map?

Springfield

2. Which city is located at point $(8, 5\frac{1}{2})$ on the map?

Shelbyville

3. Which city's location is given by an ordered pair that includes a 0?

Uniontown

4. What ordered pair describes the location of Capital City?

(7, 3)

5. If you started at (0, 0) and moved 1 unit north and 2 units east, which city would you reach?

Bryn

6. Which two cities on the map are both located 4 units to the right of (0, 0)?

Springfield and Columbus

Circle the letter of the correct answer.

7. If you started in Bart City and moved 2 units south and 2 units west, which city would you reach?
   - A Columbus
   - B Sunnydale
   - (C) Homer
   - D Bakersville

8. Starting at (0, 0), which of the following directions would lead you to Capital City?
   - (F) Go 7 units east and 3 units north.
   - G Go 5 units north and 3 units east.
   - H Go 3 unit east and 7 units north.
   - J Go 8 units east and 6 units north.

LESSON 6-6

Reading Strategies

Sequencing Directions

A **coordinate grid** is formed by a series of horizontal and vertical lines. Each point on the grid can be shown by using an **ordered pair.**

Reading a map of the city is like finding a point on a **coordinate grid.**

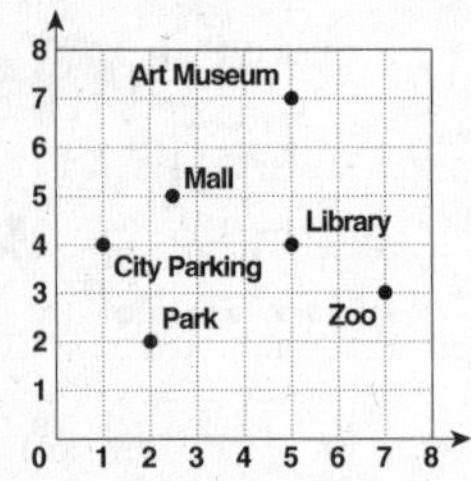

This sequence of steps helps you find points on the coordinate grid:
Step 1: Start at 0 on the coordinate grid.
Step 2: The first number tells you how far to move to the right.
Step 3: The second number tells you how far to move up.
For example, the coordinates (7, 3) mean:

Start at 0. Move right 7 units. Go up 3 units. → You're at the zoo!

Answer each question.

1. Write the sequence of steps that will get you to the library.
   Start at 0. Go right 5 units. Go up 4 units.
2. What ordered pair gives the location of the library? **(5, 4)**
3. Write the sequence of steps that will get you to the mall.
   Start at 0. Go right $2\frac{1}{2}$ units. Go up 5 units.
4. What ordered pair gives the location of the mall? **$(2\frac{1}{2}, 5)$**
5. The word *sequence* means to put things in order. Why do you think the numbers used to graph a point on the grid are called ordered pairs?
   Possible answer: because you have to follow the correct order to find points on the grid

LESSON 6-6

Puzzles, Twisters & Teasers

Griddle

In a griddle (grid riddle) the answer to the riddle is hidden within the grid. To solve the riddle, locate the letters in the grid by the given ordered pairs.

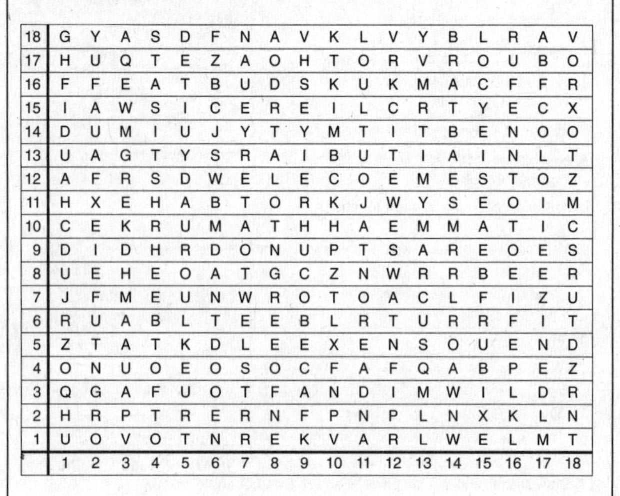

18	G	Y	A	S	D	F	N	A	V	K	L	V	Y	B	L	R	A	V
17	H	U	Q	T	E	Z	A	O	H	T	O	R	V	R	O	U	B	O
16	F	F	E	A	T	B	U	D	S	K	U	K	M	A	C	F	F	R
15	I	A	W	S	I	C	E	R	E	I	L	C	R	T	Y	E	C	X
14	D	U	M	I	U	J	Y	T	Y	M	T	I	T	B	E	N	O	O
13	U	A	G	T	Y	S	R	A	I	B	U	T	I	A	I	N	L	T
12	A	F	R	S	D	W	E	L	E	C	O	E	M	E	S	T	O	Z
11	H	X	E	H	A	B	T	O	R	K	J	W	Y	S	E	O	I	M
10	C	E	K	R	U	M	A	T	H	H	A	E	M	M	A	T	I	C
9	D	I	D	H	R	D	O	N	U	P	T	S	A	R	E	O	E	S
8	U	E	H	E	O	A	T	G	C	Z	N	W	R	R	B	E	E	R
7	J	F	M	E	U	N	W	R	O	T	O	A	C	L	F	I	Z	U
6	R	U	A	B	L	T	E	E	B	L	R	T	U	R	R	F	I	T
5	Z	T	A	T	K	D	L	E	E	X	E	N	S	O	U	E	N	D
4	O	N	U	O	E	O	S	O	C	F	A	F	Q	A	B	P	E	Z
3	Q	G	A	F	U	O	T	F	A	N	D	I	M	W	I	L	D	R
2	H	R	P	T	R	E	R	N	F	P	N	P	L	N	X	K	L	N
1	U	O	V	O	T	N	R	E	K	V	A	R	L	W	E	L	M	T
	1	2	3	4	5	6	7	8	9	10	11	12	13	14	15	16	17	18

Why did Antonio quit his job at the doughnut store?

Because he decided

T (2, 5) **O** (4, 1) **Q** (13, 4) **U** (9, 9) **I** (12, 3) **T** (18, 1)

T (2, 5) **H** (3, 8) **E** (7, 6) **H** (1, 2) **O** (4, 1) **L** (11, 15) **E** (5, 4)

B (10, 13) **U** (9, 9) **S** (7, 4) **I** (12, 3) **N** (17, 5) **E** (5, 4) **S** (15, 12) **S** (4, 15).

LESSON 6-7

Practice A

Line Graphs

Use the line graph to answer each question.

1. In which month shown on the line graph does Washington usually receive the most precipitation?
   August
2. In general, how does precipitation in Washington, D.C., change between August and October?
   It decreases.
3. In which months does the city usually receive the same amount of precipitation?
   October and December

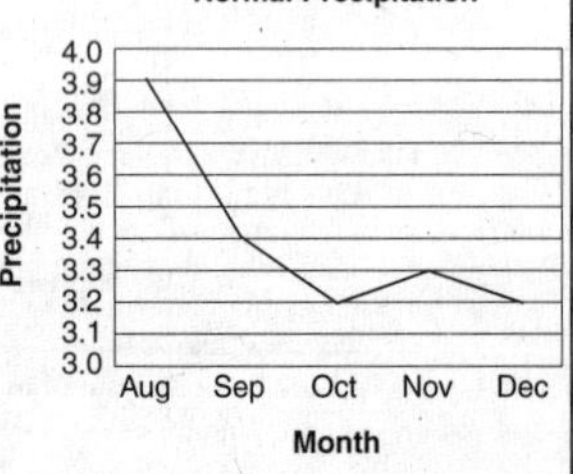

4. Use the given data to make a line graph.

Washington, D.C., Average Normal Temperatures

Month	Temperature (°F)
January	31
March	43
May	62
July	76
September	67
November	45

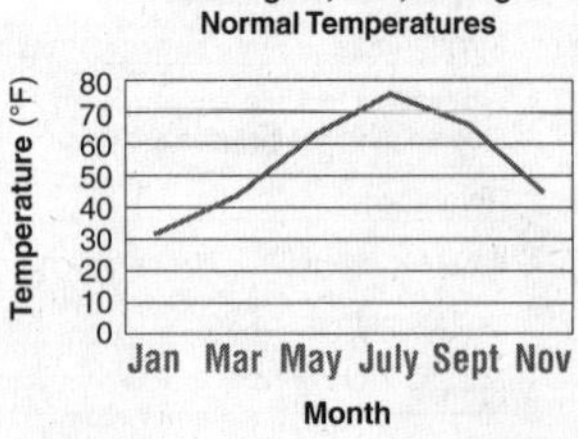

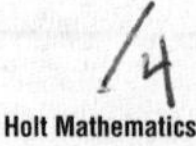

LESSON 6-7

Practice B

Line Graphs

Use the line graph to answer each question.

1. In which year were the average weekly earnings in the United States the highest?
   2000
2. In general, how did average weekly earnings in the United States change between 1970 and 2000?
   The earnings increased.
3. In which year did the average United States worker earn about $350 a week?
   1990

Average U.S. Weekly Earnings

Weekly Earnings ($): 0, 100, 200, 300, 400, 500

Year: 1970, 1980, 1990, 2000

4. Use the given data to make a line graph.

U.S. Minimum Wage

Year	Hourly Rate
1970	$1.60
1980	$3.10
1990	$3.80
2000	$5.15

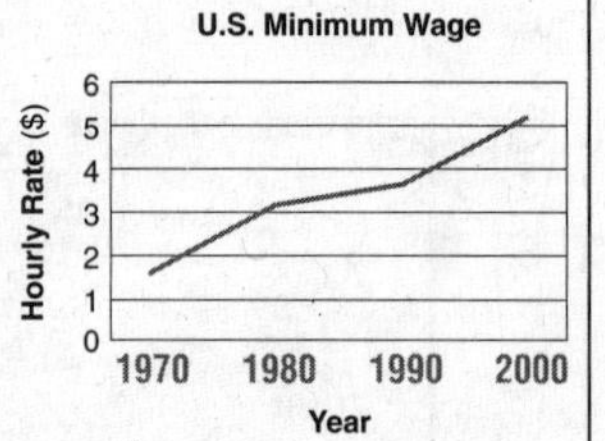

5. Between which two years shown on the graph did the U.S. minimum wage change the least?
   1980 and 1990
6. How has the hourly minimum wage changed in the U.S. since 1970?
   It has increased.

LESSON 6-7 Practice C
Line Graphs

Use the double-line graph to answer each question.

Baltimore and Philadelphia Populations

Population: 2,500,000; 2,000,000; 1,500,000; 1,000,000; 500,000; 0

Year: 1900, 1950, 2000

Key: —— Baltimore; ---- Philadelphia

1. In which year were the populations of Baltimore and Philadelphia the closest?

   1900

2. Since 1950, what has happened to the populations of both cities?

   They have decreased.

3. In which year did the cities have their highest populations?

   1950

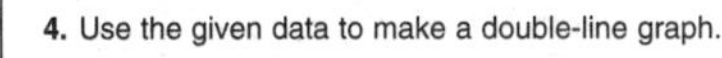

4. Use the given data to make a double-line graph.

Maryland and Pennsylvania Populations

Year	Maryland	Pennsylvania
1900	1,188,044	6,302,115
1950	2,343,001	10,498,012
2000	5,296,486	12,281,054

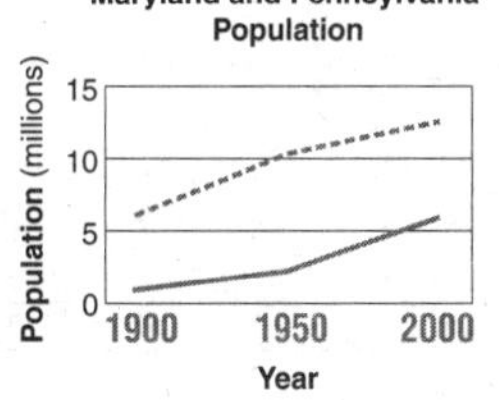

Key: —— Maryland; ---- Pennsylvania

5. In which year were the populations of Maryland and Pennsylvania the closest?

   1900

6. How did the populations of the two states compare between 1900 and 2000?

   Pennsylvania always had a higher population.

LESSON 6-7 Reteach
Line Graphs

A line graph shows change over time.

You can represent data by making a line graph.

Stock Sales (millions)

Mon	Tue	Wed	Thurs	Fri
1.5	2.0	2.25	1.75	0.5

To make a line graph, make "days" the horizontal axis and "sales" the vertical axis. Label the axes.

Then determine an appropriate scale and interval for each axis.

Think of the data in the table as ordered pairs. Mark a point for each ordered pair. Then connect the points with straight segments.

Make sure your line graph has a title.

Stock Sales

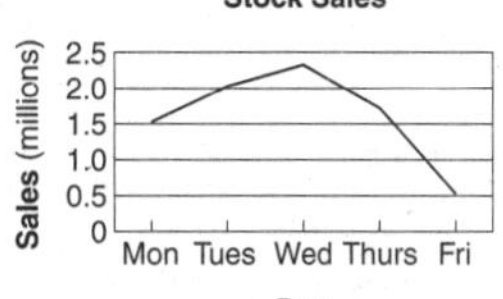

Day

Use the data in the table to make a line graph.

1. **Millie's Savings Account**

Jan	Feb	March	April	May
30	40	35	45	25

Millie's Savings Account

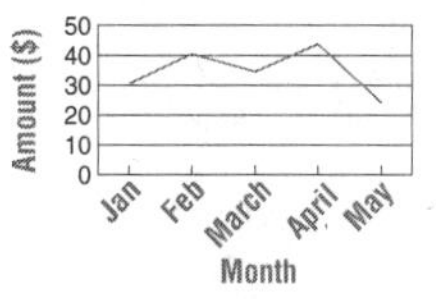

Month

LESSON 6-7 Reteach
Line Graphs (continued)

Sometimes, you need to make a double-line graph to represent data.

Stock Sales (millions)

Stock	Jan	Feb	March	April	May
A	1.5	2	2.25	1.75	0.5
B	1	2.5	2	1.5	0.75

To make a double-line graph, follow the same steps for making a line graph. Mark and connect points for each of the two sets of data you are displaying. Because there are two sets of data, make a key. Be sure to title the graph and label the axes.

Stock Sales

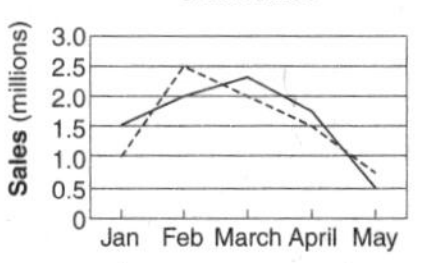

Month

Key: —— Stock A; ---- Stock B

Use the data in the table to make a double-line graph.

2. **Savings Account**

Student	Jan	Feb	March	April	May
Michael	20	10	30	25	35
Janet	30	20	35	25	45

Savings Account

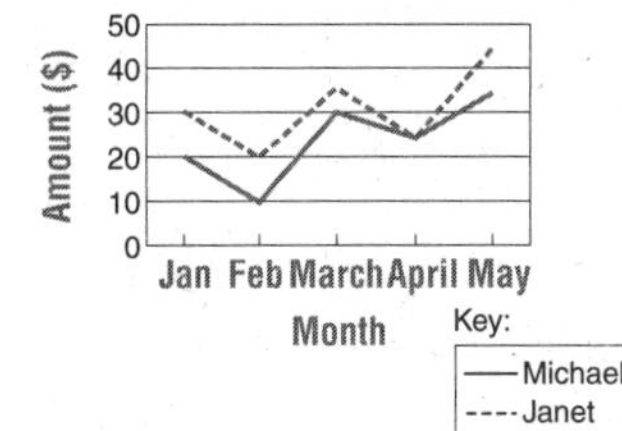

Key: —— Michael; ---- Janet

LESSON 6-7 Challenge
A Trendy Park

Because line graphs show changes over time, you can use them to make predictions based on trends, or patterns. United States park rangers make line graphs to look for trends. They count the number of people who visit their parks each month. Then the park rangers analyze the data on line graphs to look for trends and predict how many visitors to expect each month in the coming years. This data helps the rangers schedule workers and provide services for their visitors.

Great Smoky Mountains National Park in Tennessee and North Carolina receives more visitors each year than any other national park. Imagine you are a park ranger there. Use the line graph below to identify trends and make predictions about the number of visitors the park will receive in the future.

Visitors At Great Smoky Mountains National Park, 2000

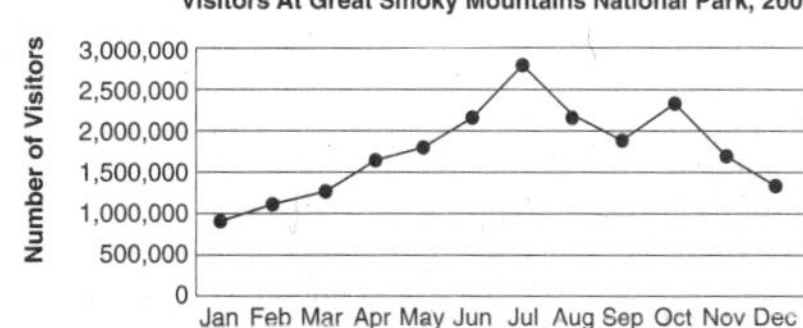

Month

Possible answers:

1. In which month next year should you plan for the most visitors at your park?

   July

2. What can you expect next year at the park between October and January?

   Each month the number of visitors will decrease.

3. You are in charge of deciding how many rangers should be scheduled to work at Great Smoky Mountains National Park each month next year. How will the number of park rangers you schedule change each month from January to June?

   Each month, I will schedule more park rangers to work than the month before.

4. Which month next year would be best for you to take time off from your park ranger job and go on your own vacation? Explain.

   January; because that is the month when the park has the least amount of visitors.

LESSON 6-7
Problem Solving
Line Graphs

Use the line graphs to answer each question.

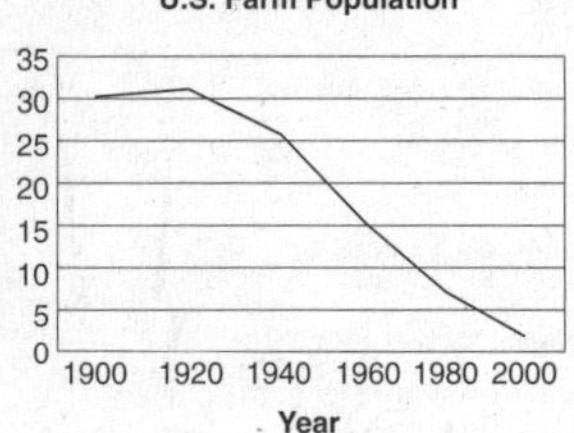

Size of U.S. Farms

Average Farm Size (acres)

500 450 400 350 300 250 200 150 100 50 0

1900 1920 1940 1960 1980 2000

Year

1. In which year was the U.S. farm population the highest? the lowest?

   1920; 2000

2. In which year was the size of the average U.S. farm the largest? the smallest?

   2000; 1900

3. In general, how has the U.S. farm population changed in the last 100 years?

   The population has decreased.

4. In general, how has the size of the average U.S. farm changed in the last 100 years?

   The average size has increased.

Circle the letter of the correct answer.

5. How many people lived on farms in the United States in 1940?
   - **A** 31 million
   - **B** 30 million
   - ⓒ 26 million
   - **D** 15 million

6. How many acres did the average farm in the United States cover in 1980?
   - **F** 150 acres
   - **G** 300 acres
   - **H** 400 acres
   - Ⓙ 426 acres

7. Between which two years did the U.S. farm population increase?
   - Ⓐ 1900 and 1920
   - **B** 1920 and 1940
   - **C** 1940 and 1960
   - **D** 1960 and 1980

8. Between which two years did the average size of farms in the United States change the least?
   - Ⓕ 1900 and 1920
   - **G** 1920 and 1940
   - **H** 1960 and 1980
   - **J** 1980 and 2000

LESSON 6-7
Reading Strategies
Reading a Graph

A **line graph** shows how data changes over a period of time. The line graph below shows a family's weekly food costs for four weeks.

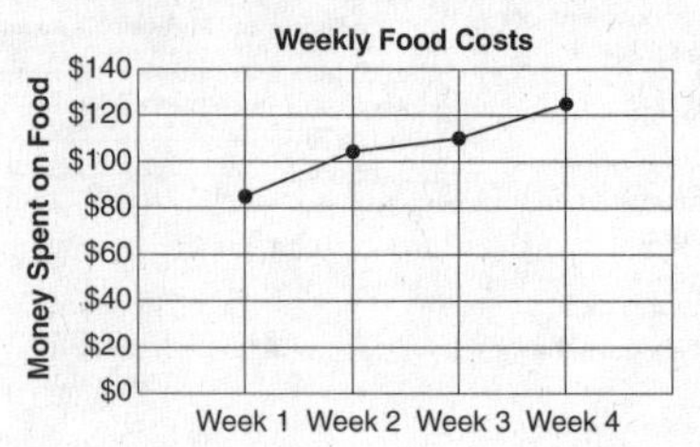

Answer the following questions about the line graph.

1. What information is located along the left side of the graph?

   number of dollars

2. By what amount do the dollars increase on the left side of the graph?

   by $20

3. What information is shown along the bottom of the graph?

   the weeks that grocery costs were tracked

Each point on the graph identifies the amount of money spent, by week.

4. How much money was spent on food in Week 1?

   $85

5. How much money was spent on food in Week 4?

   $125

6. From the line graph, what can you conclude about food costs for this family?

   Possible answer: Their food costs continue to rise from week to week.

16

LESSON 6-7
Puzzles, Twisters & Teasers
Frogs, Frogs, Frogs!

Every summer, from 1994 to 2004, three brothers would go to the lake and catch frogs. The graph shows the record they kept of their catches.

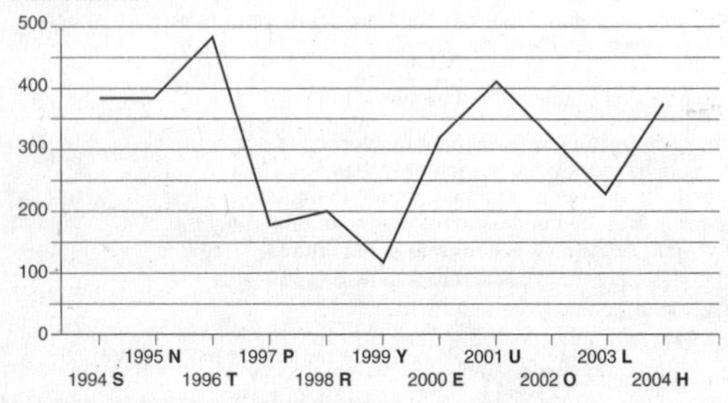

Find the year that answers the questions below and the letter beside that year. Once you have the answer, put the letter that is next to the answer over the question number in the riddle answer.

1. The year they caught about 175 frogs. 1997, P
2. The only year when they caught the same number of frogs as the previous year. 1995, N
3. The year they caught the least number of frogs. 1999, Y
4. The only year when they caught about 150 more frogs than the year before. 2004, H
5. The year when they had the biggest difference from the year before. 1997, P
6. The year they caught the second most frogs. 2001, U
7. The year when they caught less than the year before but more than the year after. 2002, O

How did the frog feel when he broke his leg?

U	N	H	O	P	P	Y
6	2	4	7	5	1	3

18

LESSON 6-8
Practice A
Misleading Graphs

Use the graph to answer each question. Possible answers are given.

1. Why is this bar graph misleading?

   Because the lower part of the vertical scale is missing, the difference in games won is exaggerated.

2. What might people believe from the misleading graph?

   Lee High School won 4 times as many games than Grant High School.

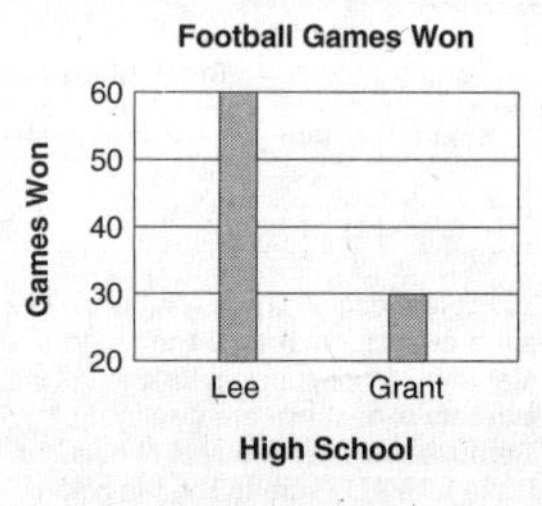

Use the graph to answer each question.

3. Why is this line graph misleading?

   Because there is a break in the vertical scale, the differences in tickets sold are exaggerated.

4. What might people believe from the misleading graph?

   About 4 times more tickets were sold in January than in September.

A

LESSON 6-8

Practice B
Misleading Graphs

Use the graph to answer each question. Possible answers are given.

1. Why is this bar graph misleading?

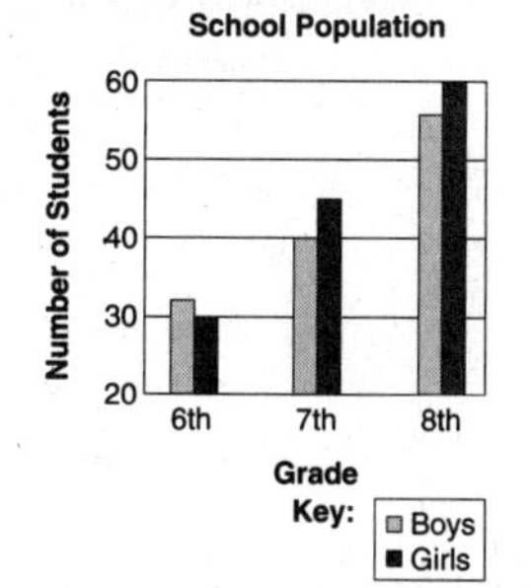

Because the lower part of the vertical scale is missing, the differences in grades are exaggerated.

2. What might people believe from the misleading graph?

There are 4 times as many students in the 8th grade than the 6th grade.

Use the graph to answer each question.

3. Why is this line graph misleading?

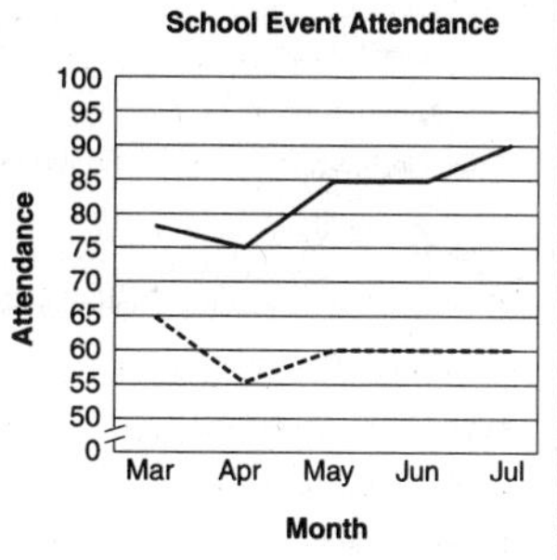

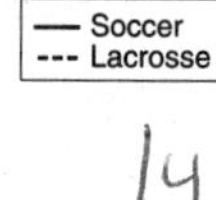

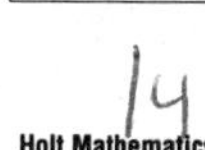

Because there is a break in the vertical scale, the differences in attendance seem greater than they really are.

4. What might people believe from the misleading graph?

In some months, 3 times more people attended soccer games than lacrosse games.

LESSON 6-8

Practice C
Misleading Graphs

Use the graph to answer each question. Possible answers are given.

1. Why is this bar graph misleading?

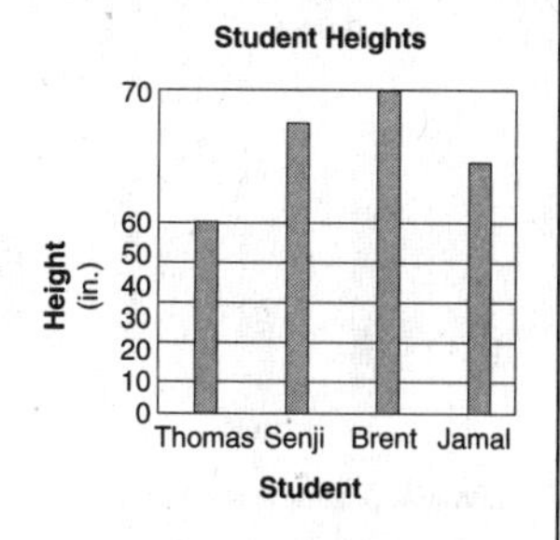

Because the vertical scale intervals are unequal, the differences in height are exaggerated.

2. What might people believe from the misleading graph?

That Brent is twice as tall as Thomas, when he really is only 10 inches taller.

Use the graphs to answer each question.

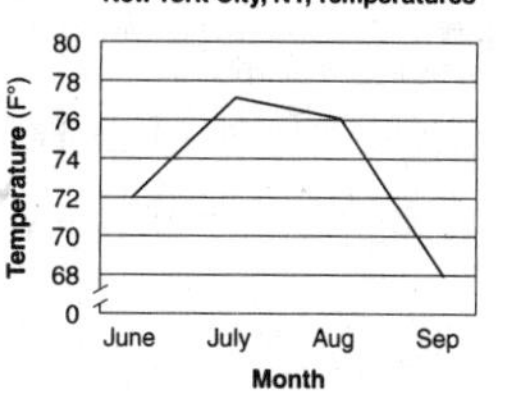

Albany, NY, Temperatures (Temperature (F°): 0–80; Month: June, July, Aug, Sep)

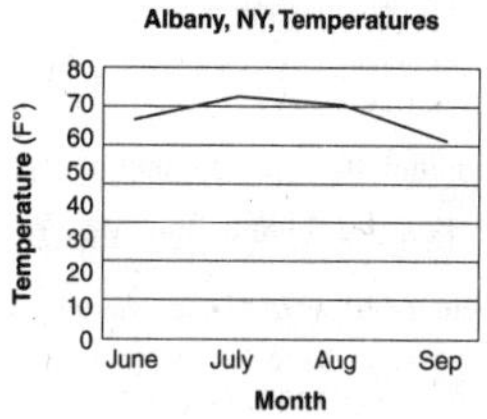

3. Why are these line graphs misleading?

Because the graphs' scales are different, and NYC's graph has a break in the scale, the differences in the two cities' temperatures are exaggerated.

4. What might people believe from the misleading graphs?

Albany has higher summer temperatures than NYC; summer temperatures in NYC vary more than in Albany.

LESSON 6-8

Reteach
Misleading Graphs

Graphs are often made to influence you. When you look at a graph, you need to figure out if the graph is accurate or if it is misleading.

Look at the graph below.

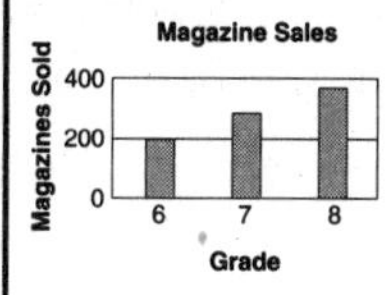

The graph is misleading because the intervals for the scale are so great. When you first look at the graph it appears that each grade sold about the same number of magazines.

Look at each graph. Then explain why each graph is misleading.

1.

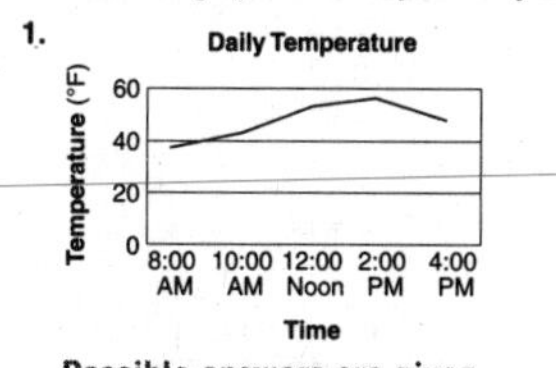

Possible answers are given.

The line graph is misleading because the intervals on the scale are so great. The graph leads you to think that the temperature changed very little throughout the day.

2.

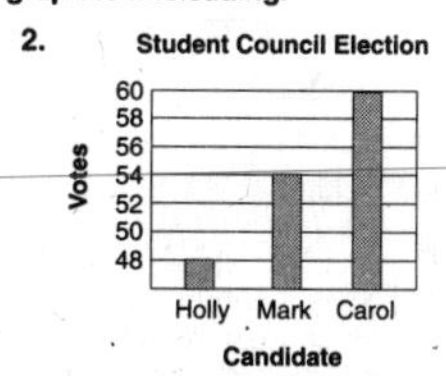

The bar graph is misleading because of the scale. The graph leads you to believe that the winner won the election by a large margin when it was actually a close race.

LESSON 6-8

Challenge
Graph Detective

You are a police detective in Capital City. A gang of criminals there is distributing misleading graphs to convince people that your city does not need to hire more police officers. It's your job to catch these graph crooks.

Search the graphs below for evidence of misleading displays of data. Then use your detective skills to explain why each graph is misleading.

Capital City Crime Rate (Number of Crimes Reported (hundreds): 0–30; Year: 1970, 1980, 1990, 2000)

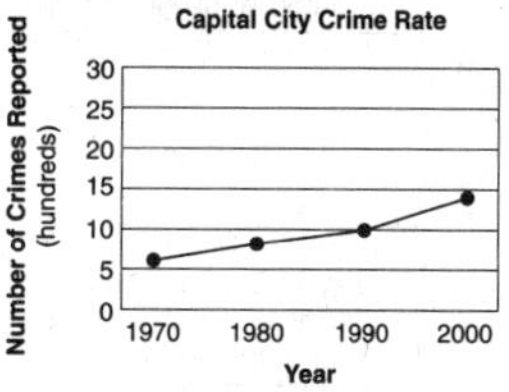

Possible answers:

Why is this line graph misleading?

The intervals on the y-axis are so large, it looks like the crime rate has changed very little.

What might people believe from this misleading graph?

The city's crime rate is low and has not changed very much since 1970, so the city does not need to hire more police officers.

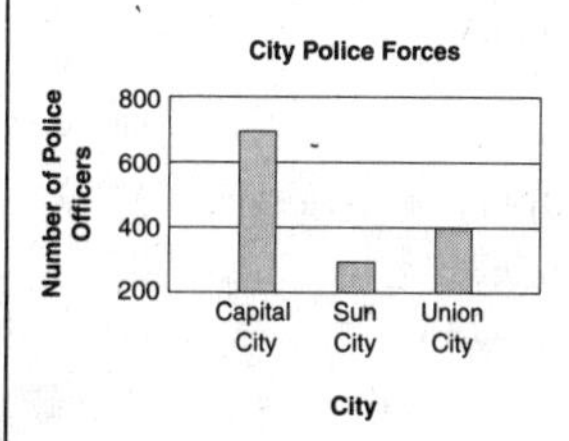

Why is this bar graph is misleading?

Because the lower part of the vertical scale is missing, the differences in the cities' police forces are exaggerated.

What might people believe from this misleading graph?

Capital City has 5 times as many police officers as Sun City and $2\frac{1}{2}$ times as many as Union City, so Capital City does not need to hire more police officers.

LESSON 6-8
Problem Solving
Misleading Graphs

Use the graphs to answer each question. Possible answers:

Graph A — Sales ($ millions): 0, 3, 6; Creamy Bars, Crispy Bars

Graph B — Sales ($ millions): 3, 4, 5; Creamy Bars, Crispy Bars

Graph C — Sales ($ millions): 0, 1, 2, 3, 4, 5, 6; Creamy Bars, Crispy Bars

1. Why is Graph A misleading?

   The vertical scale intervals are not equal, which makes the data look closer than it actually is.

2. Why is Graph B misleading?

   The lower part of the vertical scale is missing; differences in sales are exaggerated.

3. What might people believe from reading Graph A?

   About the same number of Crispy Bars and Creamy Bars were sold.

4. What might people believe from reading Graph B?

   Creamy Bar sales were twice the sales of Crispy Bar sales.

Circle the letter of the correct answer.

5. Which of the following information is different on all three graphs above?
   - (A) the vertical scale (circled)
   - B the Crispy Bars sales data
   - C the Creamy Bars sales data
   - D the horizontal scale

6. Which of the following is a way that graphs can be misleading?
   - F breaks in scales
   - G uneven scales
   - H missing parts of scales
   - (J) all of the above (circled)

7. Which graph do you think was made by the company that sells Crispy Bars?
   - (A) Graph A (circled)
   - B Graph B
   - C Graph C
   - D all of the graphs

8. If you were writing a newspaper article about candy bar sales, which graph would be best to use?
   - F Graph A
   - G Graph B
   - (H) Graph C (circled)
   - J all of the above

LESSON 6-8
Reading Strategies
Compare and Contrast

A graph can be misleading if the scale of the graph does not start at 0. Compare Graph A and Graph B.

Graph A — Favorite Places to Visit — Number of Votes: 0, 10, 20, 30, 40, 50; Zoo, Beach, Park, Museum

Graph B — Favorite Places to Visit — Number of Votes: 20, 30, 40, 50, 60, 70; Zoo, Beach, Park, Museum

Answer these questions about the graphs.

1. What are the titles of Graph A and Graph B?

   Favorite Places to Visit

2. What do the numbers on the left side of the graph show?

   the number of votes

3. By how much do the number of votes increase along the left side of the graph?

   by 10

4. What do the bars on the graph stand for?

   number of votes for that place

5. What is the first number shown along the left side of Graph A? 0

6. What is the first number shown along the left side of Graph B? 20

7. Compare the bars on Graph A to the bars on Graph B.

   The bars on Graph A are longer than the bars on Graph B.

8. Which graph is misleading?

   Graph B

/8

LESSON 6-8
Puzzles, Twisters & Teasers
Doctor, Doctor

Two doctors are advertising in the paper. Here are their newspaper advertisements.

Graph #1

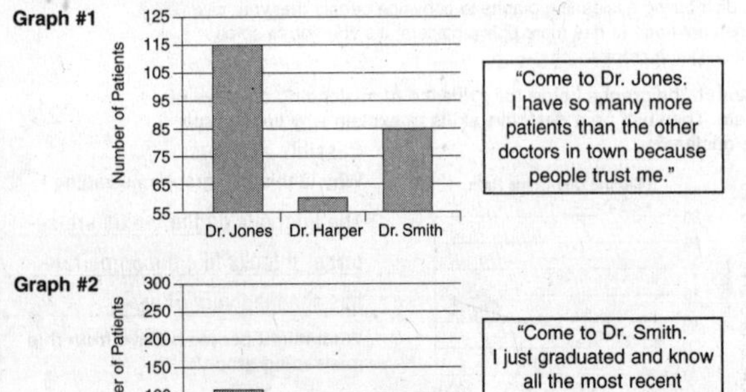

Graph #2

Number of Patients: 0, 50, 100, 150, 200, 250, 300; Dr. Jones, Dr. Harper, Dr. Smith

"Come to Dr. Smith. I just graduated and know all the most recent medical procedures."

Answer the questions to solve the riddle:

1. In which graph does it look like Doctor Jones has about the same number of patients as other doctors? 2 P
2. By what number do the numbers increase on the *y*-axis in graph #2? 50 N
3. By what number do the numbers increase on the *y*-axis in graph #1? 10 A
4. What is the highest number on the *y*-axis on graph #2? 300 I
5. About how many patients does Doctor Jones have? 115 T
6. About how many more patients does Doctor Jones see than Dr. Smith? 30 E
7. If you were Dr. Jones, which graph would you want people to see? 1 T

A man runs into the doctor's office and says, "Hey Doc, I think I'm shrinking. Do something right away!"

The doctor replies:

"YOU'LL JUST HAVE TO BE A LITTLE P A T I E N T."
2 10 115 300 30 50 1

LESSON 6-9
Practice A
Stem-and-Leaf Plots

Complete each activity and answer each question.

1. Use the data in the table to complete the stem-and-leaf plot below.

Daily Low Temperatures (°F)	16	21	15	27	30	25

Check students' stem-and-leaf plots.

Daily Low Temperature

Stem	Leaves
1	5 6
2	1 5 7
3	0

Key: 1 | 6 = 16°F

Find each value of the data.

Stem	Leaves
1	0 5
2	3
3	7
4	5

Key: 2 | 4 = 24

2. smallest value 10
3. largest value 45
4. mean 26
5. median 23
6. mode none
7. range 35

8. In the stem-and-leaf plot for Exercises 2–7, which digit was used for the stems? for the leaves?

   stems: tens; leaves: ones

9. Look at the stem-and-leaf plot you made for Exercise 1. What are the smallest and largest values in the data set?

   smallest: 15; largest: 30

LESSON 6-9 Practice B
Stem-and-Leaf Plots

Complete each activity and answer the questions.

1. Use the data in the table to complete the stem-and-leaf plot below.

Richmond, Virginia, Monthly Normal Temperatures (°F)											
Jan	Feb	Mar	April	May	June	July	Aug	Sep	Oct	Nov	Dec
37	39	48	57	74	78	77	76	70	59	50	40

Stem	Leaves
3	7 9
4	0 8
5	0 7 9
6	
7	0 4 6 7 8

Key: 1 | 2 = 12°F

Find each value of the data.

Stem	Leaves
6	1 4
7	1 6
8	2 2
9	0 1 8

Key: 6 | 5 = 65

2. least value 61
3. greatest value 98
4. mean 79.4
5. median 82
6. mode 82
7. range 37
8. Look at the stem-and-leaf plot you made for Exercise 1. How many months in Richmond have a normal temperature above 70°F?
   4 months
9. How would you display a data value of 100 on the stem-and-leaf plot above?
   Use 10 for the stem and 0 for the leaf.

LESSON 6-9 Practice C
Stem-and-Leaf Plots

Complete each activity and answer the questions.

1. Use the data in the table to make a stem-and-leaf plot.

Books Read by Read-A-Thon Participants									
50	19	24	45	44	12	32	19	38	43
35	40	15	19	26	30	28	40	12	18

Stem	Leaves
1	2 2 5 8 9 9 9
2	4 6 8
3	0 2 5 8
4	0 0 3 4 5
5	0

Key: 1 | 2 = 12

Find each value of the data.

Stem	Leaves
1	2 5 8 9
2	0 4 5 6 8 8
3	2 6
5	0 2

Key: 3 | 3 = 33

2. least value 12
3. greatest value 52
4. mean 27.5
5. median 25.5
6. mode 28
7. range 40
8. Look at the stem-and-leaf plot you made for Exercise 1. How many students read more than 40 books during the read-a-thon?
   4 students
9. How would you display a data value of 5 on the stem-and-leaf plot above? What would be the mean of this new data set?
   Use 0 for the stem and 5 for the leaf; mean: 26

CHAPTER 6-9 Reteach
Stem-and-Leaf Plots

You can use place value to make a stem-and-leaf plot.

Points Earned in Games During Basketball Season						
27	16	34	29	48	12	33
20	18	42	51	27	32	41

Write the numbers in order from least to greatest.

12 16 18 20 27 27 29 32 33 34 41 42 48 51

List the tens digits in order from least to greatest in the first, or stem, column. Then, for each tens digit, record the ones digit for each data value in order from least to greatest in the second, or leaves, column.

Points Earned

Stem	Leaves
1	2 6 8
2	0 7 7 9
3	2 3 4
4	1 2 8
5	1

Key: 1 | 2 = 12

Make sure your graph has a title and a key.

Use the data to make a stem-and-leaf plot.

1.

Valerie's Test Scores				
62	84	93	88	89
76	68	81	91	88

Valerie's Test Scores

Stem	Leaves
6	2 8
7	6
8	1 4 8 8 9
9	1 3

Key: 6 | 2 = 62

2. What is the range? The range is 31 points.
3. What is the median? The median is 86 points.
4. What is the mode? The mode is 88 points.

LESSON 6-9 Challenge
A Plot of Trees

As part of your job with the National Forest System, you must compile a report on the tallest trees in the United States. You have already collected the data and displayed it on the bar graph below. To complete the report, you need to make a stem-and-leaf plot of the data. Use the hundreds digits of the data for your stems. Then analyze the data.

National Forest System Report, 2000

Tallest Trees in the U.S.

Stem	Leaves
1	78 94
2	06 19 27 32 72 75 81
3	21

Key: 1 | 78 = 178 feet

Data Analysis

Range of heights: 143 feet
Mean height: 240.5 feet
Median height: 229.5 feet
Mode height: none

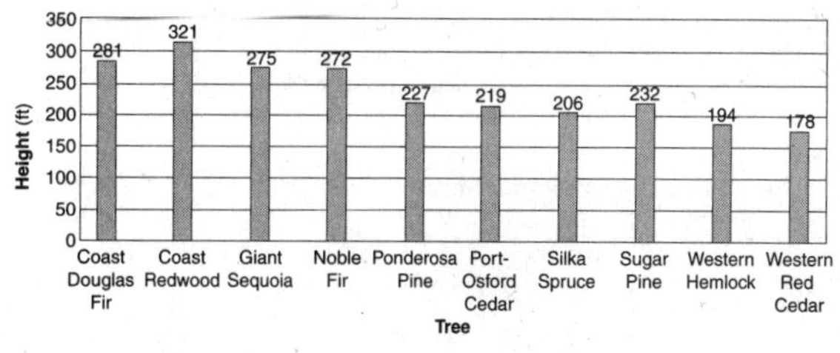

LESSON 6-9
Problem Solving
Stem-and-Leaf Plots

Use the Texas stem-and-leaf plots to answer each question.

Dallas Normal Monthly Temperatures

Stem	Leaves
4	3 7 8
5	6 7
6	6 7
7	3 7
8	1 5 5

Key: 4 | 3 = 43°F

Houston Normal Monthly Temperatures

Stem	Leaves
5	0 4 4
6	1 1 8
7	0 5 8
8	0 2 3

Key: 5 | 0 = 50°F

1. Which city's temperature data has a mode of 85°F?

   Dallas

2. Which city's temperature data has a range of 33°F?

   Houston

3. Which city has the lowest data value? What is that value?

   Dallas; 43°F

4. Which city has the highest data value? What is that value?

   Dallas; 85°F

Circle the letter of the correct answer.

5. Which city's temperature data has a mean of 68°F?
   - A Dallas
   - (B) Houston
   - C both Dallas and Houston
   - D neither Dallas nor Houston

6. Which city's temperature data has a median of 69°F?
   - F Dallas
   - (G) Houston
   - H both Dallas and Houston
   - J neither Dallas nor Houston

7. What do the data values 54°F and 61°F represent for the plots above?
   - A the ranges of normal temperatures in Dallas and Houston
   - (B) the mode of normal temperatures for Houston
   - C the mean and median normal temperatures for Dallas
   - D the lowest normal temperatures for Dallas and Houston

8. Which of the following would be the best way to display the Dallas and Houston temperature data?
   - (F) on a line graph
   - G in a tally table
   - H on a bar graph
   - J on a coordinate plane

LESSON 6-9
Reading Strategies
Use a Graphic Organizer

Below is a list of high temperatures during a two-week period in Austin, Texas.

75 78 63 79 74 73 83 72 85 62 84 65 68 81

Making a table is one way to organize the temperature data so it is easier to understand.

63	75	83
62	78	85
65	79	84
68	74	81
	72	
	73	

Answer each question about the table.

1. How were the temperatures organized in the table?

   Possible answer: Temperatures in the 60's were put in one column, temperatures in the 70's in another, and temperatures in the 80's in another.

2. How many days was the temperature in the 70's? 6 days

3. How many days was the temperature above 80? 4 days

4. Complete: Temperatures were mostly in the 70's during this two-week period.

5. Put the temperatures with 8 in the tens place in order from least to greatest.

   81, 83, 84, 85

6. What was the range of high temperatures during the two-week period?

   from 62 to 85, or 23 degrees

7. How did organizing the temperatures help you answer the questions above?

   Possible answer: The numbers were grouped together by tens, which made it easier to find answers that were in the 60's, 70's or 80's.

LESSON 6-9
Puzzles, Twisters & Teasers
It's a Leaf

Arrange the test scores below in a stem-and-leaf plot from least to greatest. After organizing the data in the plot, count and record the number of leaves per stem. Then find the letter matching the number of leaves. Solve the riddle by arranging the number of leaves from least to greatest.

Test Scores: 56, 70, 100, 65, 48, 92, 84, 95, 97, 100, 68, 75, 81, 85
59, 92, 96, 66, 75, 83, 66, 72, 85, 93, 73, 95, 100, 87

Stem	Leaves	Number of Leaves	Letter
4	8	1	Q
5	6 9	2	U
6	5 6 6 8	4	C
7	0 2 3 5 5	5	K
8	1 3 4 5 5 7	6	L
9	2 2 3 5 5 6 7	7	Y
10	0 0 0	3	I

Letters

A = 8	D = 10	I = 3	L = 6	S = 13
B = 11	E = 0	J = 9	N = 12	U = 2
C = 4	G = 14	K = 5	Q = 1	Y = 7

How do hikers dress on cold mornings?

Q U I C K L Y

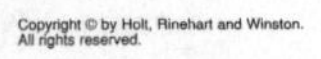

LESSON 6-10
Practice A
Choosing an Appropriate Display

1. The table shows the average high temperatures in San Diego for six months of one year. Which graph would be more appropriate to show the data—a line graph or a bar graph? Draw the more appropriate graph.

Month	Mar	June	Aug	Sep	Oct	Dec
Average High Temperature (°F)	65	70	77	76	73	64

Average High Temperatures in San Diego

Temperature (F)

Mar June Aug Sep Oct Dec

Month

2. The table shows the results of a survey about students' favorite snack. Which graph would be more appropriate to show the data—a line graph or a bar graph? Draw the more appropriate graph.

Type of Snack	Popcorn	Fruit	Cheese	Pretzels
Number of Votes	16	6	2	9

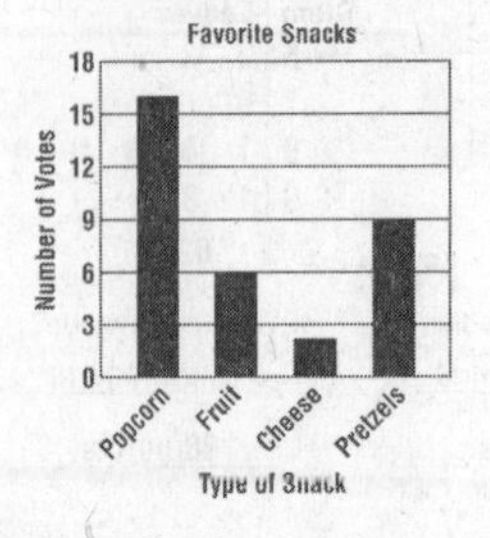

LESSON 6-10 **Practice B**

Choosing an Appropriate Display

1. The table shows the heights of the 6 tallest buildings in the world. Which graph would be more appropriate to show the data—a line graph or a bar graph? Draw the more appropriate graph.

Building	Sears Tower	CITIC Plaza	Petronas Tower I	Petronas Tower II	Jin Mao Building	Two Finance Center
Height (ft)	1,450	1,283	1,483	1,483	1,381	1,352

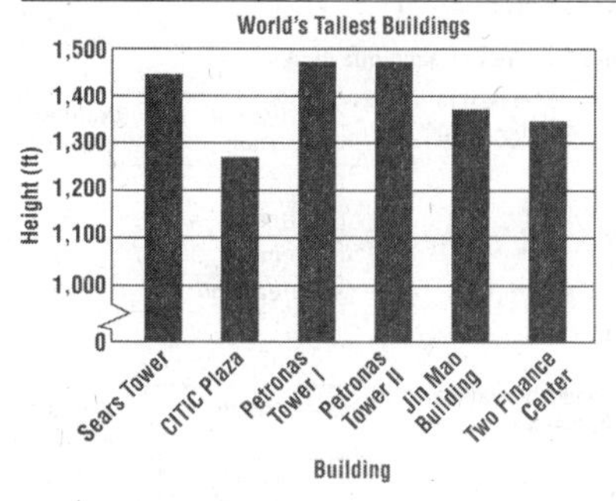

2. The table shows the test scores of some sixth-grade students. Which graph would be more appropriate to show the data—a stem-and-leaf plot or a line graph? Draw the more appropriate graph.

Test Scores												
62	78	81	66	96	88	81	77	90	88	60	99	90

Test Scores

Stems	Leaves
6	0 2 6
7	7 8
8	1 1 8 8
9	0 0 6 9

Key 6|0 = 60

12

LESSON 6-10 **Practice C**

Choosing an Appropriate Display

1. The table shows the populations of the 6 largest Native American tribes in the United State. Which graph would be more appropriate to show the data—a line graph or a bar graph? Draw the more appropriate graph.

Tribe	Navajo	Sioux	Chippewa	Cherokee	Choctaw	Latin American Indian
Population	269,202	108,272	105,907	281,069	87,349	104,354

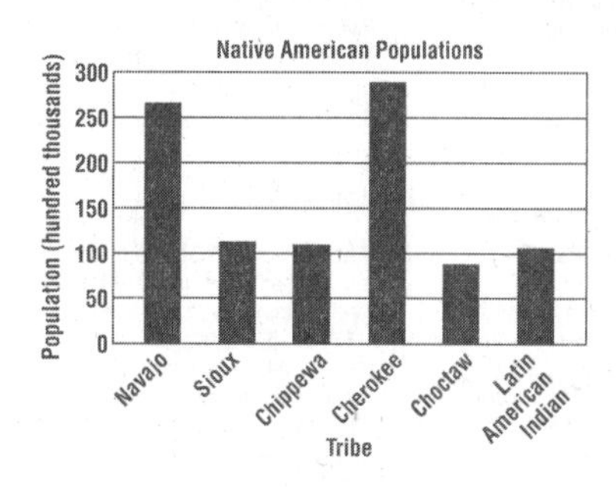

2. The table shows the heights of some sixth-grade students. Make a graph of the data, using the most appropriate way to display it.

Heights of Sixth-Graders (in.)												
52	55	60	53	51	54	55	51	53	50	50	52	51

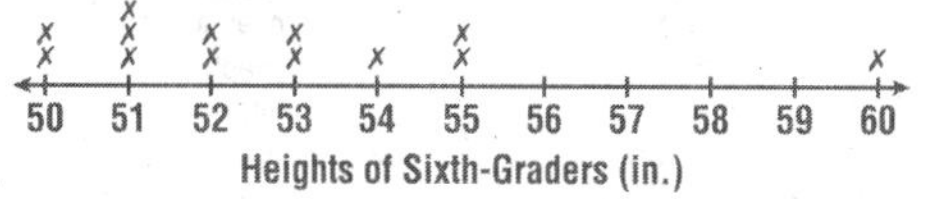

CHAPTER 6-10 **Reteach**

Choosing an Appropriate Display

You can choose the best way to display data by thinking about the data you want to show.

Use a line plot to show the frequency of data on a number line.

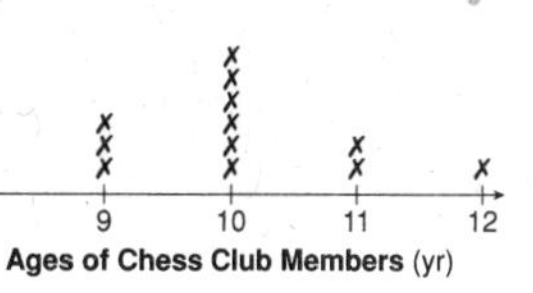

Use a line graph to show changes in data over time.

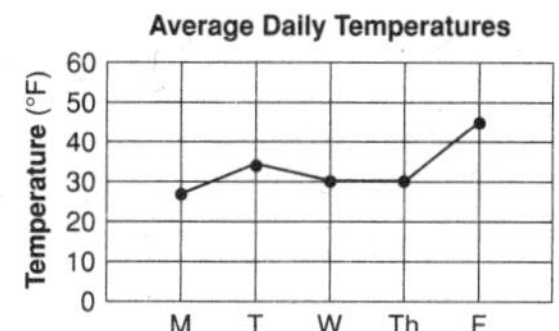

Use a bar graph to display data in separate categories.

Books Read

Number of Books: 0, 2, 4, 6, 8, 10

Al, Mia, Ray, Jen

Person

Use a stem-and-leaf plot to show how often data values occur and how they are distributed.

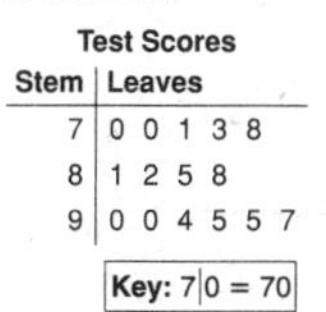

Display the data. Use the most appropriate type of graph.

Month	Mar	Apr	May	June	July
Average Temperature (°C)	12	18	20	23	25

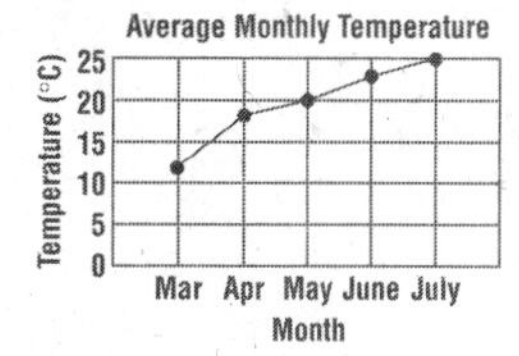

LESSON 6-10 **Challenge**

Graphing the News

Your job as a television news reporter is to make data displays for use on each evening's newscast. Tell the most appropriate type of graph to use for each kind of data described below. **Sample answers are given.**

- the daily average temperature for five days **line graph**
- the six most popular movies last month **bar graph**
- the ages of twenty world leaders **line plot**
- the amount of money collected by thirty charity organizations **stem-and-leaf plot**

Choose one of the kinds of data above and write a data set for it. Then prepare the graph for this evening's news telecast.

Check that students' data is displayed in the most appropriate way.

16

LESSON **6-10**

Problem Solving

Choosing an Appropriate Display

1. Write *line plot, stem-and-leaf plot, line graph,* or *bar graph* to describe the most appropriate way to show the height of a sunflower plant every week for one month.

   line graph

2. Write *line plot, stem-and-leaf plot, line graph,* or *bar graph* to describe the most appropriate way to show the number of votes received by each candidate running for class president

   bar graph

3. Write *line plot, stem-and-leaf plot, line graph,* or *bar graph* to describe the most appropriate way to show the test scores each student received on a math quiz.

   stem-and-leaf plot

4. Write *line plot, stem-and-leaf plot, line graph,* or *bar graph* to describe the most appropriate way to show the average time spent sleeping per day by 30 sixth-grade students.

   line plot

Circle the letter of the correct answer.

5. People leaving a restaurant were asked how much they spent for lunch. Here are the results of the survey to the nearest dollar: $8, $7, $9, $7, $10, $5, $8, $8, $12, $8. Which type of graph would be most appropriate to show the data?

   A bar graph
   B line graph
   (C) line plot
   D stem-and-leaf plot

6. People leaving a movie theater were asked their age. Here are the results of the survey to the nearest year: 12, 11, 13, 15, 22, 31, 40, 12, 17, 20, 33, 16, 12, 24, 19. Which type of graph would be most appropriate to show the data?

   F bar graph
   G line graph
   H line plot
   (J) stem-and-leaf plot

7. What is the median amount of money spent on lunch in Exercise 5?

   A $7
   (B) $8
   C $9
   D $12

8. What is the median age of the movie-goers in Exercise 6?

   F 15
   G 16
   (H) 17
   J 19

LESSON **6-10**

Reading Strategies

Understand Vocabulary

A **line plot** is a number line with marks or dots that show frequency of data.

A **bar graph** is a graph that uses vertical or horizontal bars to display data in discrete categories.

A **line graph** is a graph that uses line segments to show how data changes over time.

A **stem-and-leaf plot** is a graph used to organize and display data so that the frequencies can be compared.

Use the definitions to help you answer each question.

1. Which type of display would be best to use to show the data in the table at right?

   bar graph

Type of Club	Number of Members
chess	12
crafts	23
music	14
reading	19

2. Which type of display would be best to use to show the data in the table at right?

   line graph

Month	Average Rainfall (in.)
February	1.2
March	2.4
April	3.8
May	2.1

3. Which type of display would be best to use to show the data in the table at right?

   line plot

Ages of Club Members					
11	12	11	10	11	12
13	10	14	11	10	9
10	11	12	12	11	11
11	11	10	11	10	10

LESSON **6-10**

Puzzles, Twisters & Teasers

Way to Display!

Yoko, Ken, Olivia, and Larry want to make data displays. Use the following descriptions to match the student to the best type of display for his or her data.

- Yoko wants to display the number of kilometers several people ran in last week's run for charity.
- Ken wants to display the History Museum attendance over a period of six months.
- Olivia wants to display the science test scores of all of her classmates.
- Larry wants to display the results of a survey he took about some friends' favorite weekend activities.

	1 LINE PLOT	2 STEM-AND-LEAF PLOT	3 BAR GRAPH	4 LINE GRAPH
Ken				■
Olivia		■		
Yoko	■			
Larry			■	

To solve the riddle, fill the spaces above each number with the first letter of the name of the person who will make that display number.

What do you call a funny book about eggs?

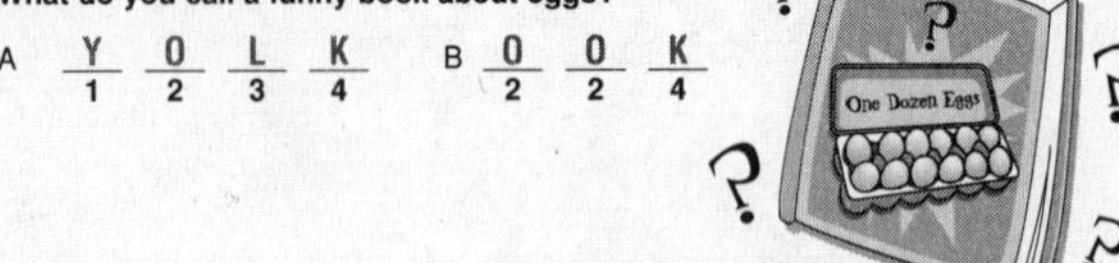

A Y O L K B O O K
(1 2 3 4 — 2 2 4)